Markus Weber

Reconstruction of the Galactic Dark Matter Density

Markus Weber

Reconstruction of the Galactic Dark Matter Density

from Astronomical Observations and Diffuse Galactic Gamma Rays

Südwestdeutscher Verlag für Hochschulschriften

Impressum/Imprint (nur für Deutschland/ only for Germany)
Bibliografische Information der Deutschen Nationalbibliothek: Die Deutsche Nationalbibliothek verzeichnet diese Publikation in der Deutschen Nationalbibliografie; detaillierte bibliografische Daten sind im Internet über http://dnb.d-nb.de abrufbar.

Alle in diesem Buch genannten Marken und Produktnamen unterliegen warenzeichen-, marken- oder patentrechtlichem Schutz bzw. sind Warenzeichen oder eingetragene Warenzeichen der jeweiligen Inhaber. Die Wiedergabe von Marken, Produktnamen, Gebrauchsnamen, Handelsnamen, Warenbezeichnungen u.s.w. in diesem Werk berechtigt auch ohne besondere Kennzeichnung nicht zu der Annahme, dass solche Namen im Sinne der Warenzeichen- und Markenschutzgesetzgebung als frei zu betrachten wären und daher von jedermann benutzt werden dürften.

Verlag: Südwestdeutscher Verlag für Hochschulschriften GmbH & Co. KG
Dudweiler Landstr. 99, 66123 Saarbrücken, Deutschland
Telefon +49 681 37 20 271-1, Telefax +49 681 37 20 271-0
Email: info@svh-verlag.de
Zugl.: Karlsruhe, TU, Diss., 2010

Herstellung in Deutschland:
Schaltungsdienst Lange o.H.G., Berlin
Books on Demand GmbH, Norderstedt
Reha GmbH, Saarbrücken
Amazon Distribution GmbH, Leipzig
ISBN: 978-3-8381-2385-1

Imprint (only for USA, GB)
Bibliographic information published by the Deutsche Nationalbibliothek: The Deutsche Nationalbibliothek lists this publication in the Deutsche Nationalbibliografie; detailed bibliographic data are available in the Internet at http://dnb.d-nb.de.

Any brand names and product names mentioned in this book are subject to trademark, brand or patent protection and are trademarks or registered trademarks of their respective holders. The use of brand names, product names, common names, trade names, product descriptions etc. even without a particular marking in this works is in no way to be construed to mean that such names may be regarded as unrestricted in respect of trademark and brand protection legislation and could thus be used by anyone.

Publisher: Südwestdeutscher Verlag für Hochschulschriften GmbH & Co. KG
Dudweiler Landstr. 99, 66123 Saarbrücken, Germany
Phone +49 681 37 20 271-1, Fax +49 681 37 20 271-0
Email: info@svh-verlag.de

Printed in the U.S.A.
Printed in the U.K. by (see last page)
ISBN: 978-3-8381-2385-1

Copyright © 2011 by the author and Südwestdeutscher Verlag für Hochschulschriften GmbH & Co. KG and licensors
All rights reserved. Saarbrücken 2011

Contents

List of Figures   iii

List of Tables   vii

1   **Introduction**   1

2   **Theoretical framework**   5
    2.1   The Cosmological Principle   5
    2.2   The Standard Model of Cosmology   7
        2.2.1   Hubble Expansion   12
        2.2.2   Cosmic Microwave Background Radiation   13
        2.2.3   Primordial Nucleosynthesis   20
    2.3   Inflation   22
    2.4   Structure formation   25
        2.4.1   Jeans instability   27
    2.5   The Milky Way   29
        2.5.1   Coordinate systems in the Milky Way   32
    2.6   Dark Matter   34
        2.6.1   Dark Matter Candidates   34
        2.6.2   Relic Density   38
        2.6.3   Annihilation of Dark Matter particles   39
    2.7   Background Processes   45

3   **Constraint on the Dark Matter density distribution from astronomical observations**   49
    3.1   Astronomical observations   49
        3.1.1   Rotation velocity and Galactocentric distance to the Sun   50

		3.1.2 Rotation curve of the Milky Way	52
		3.1.3 Mass of the Galaxy	56
		3.1.4 Local surface density and Oort limit	57
	3.2	Parametrisation of the density distribution of the Milky Way	60
		3.2.1 Luminous matter density	60
		3.2.2 Dark Matter density	61
	3.3	Results	62
		3.3.1 Local DM density	62
		3.3.2 Ringlike Dark Matter substructure	69
	3.4	Discussion	71

4 Constraints on the Dark Matter density distribution from Galactic gamma rays — 75

		4.0.1 The EGRET experiment	75
	4.1	Reconstruction of the Dark Matter density distribution	78
	4.2	Results	81
		4.2.1 Energy spectrum of the diffuse Galactic gamma radiation	81
		4.2.2 Spatial distribution of the diffuse Galactic gamma radiation	82
		4.2.3 Clumpiness of the Dark Matter rings	96
	4.3	Discussion	99
	4.4	Recent and future gamma ray observations	100
		4.4.1 Future indirect DM searches	106

5 Summary and Outlook — 111

A GALPROP datacard — 115

B Rotation velocities — 119

C Single Profile plots — 131

D Double Profile plots — 137

Bibliography — 155

List of Figures

2.1	Map of the galaxy distribution	6
2.2	Visualisation of spatial curvatures	11
2.3	Hubble's velocity-distance diagram	14
2.4	WMAP map of the CMBR temperature anisotropies	15
2.5	Angular power spectrum of the CMBR brightness fluctuations	16
2.6	Combination of the measurement of the CMBR, distance measurements with Type Ia supernovae and baryon acoustic observations.	18
2.7	Predicted primordial abundances of light elements	22
2.8	Schematic figure of the temperature potential $V_T(\phi)$	23
2.9	Filamentary structure of mass distribution of the Universe	28
2.10	Local Group and Galactic satellites	30
2.11	Illustration of the spiral structure of the Milky Way	32
2.12	Coordinate systems in the Milky Way	33
2.13	Solution of the Boltzmann equation	40
2.14	Main neutralino annihilation channels	40
2.15	Survival probability for DM clumps	44
3.1	Calculation of the Oort Constants	50
3.2	Velocity distribution in the Galactic disc of the MW.	55
3.3	Radial dependence of the different halo profiles	63
3.4	Correlation between $\rho_{\odot,\mathrm{DM}}$ and $\mathrm{M_{tot}}$	64
3.5	Total Galactic mass and rotation curve for different DM halo profiles	67
3.6	Velocity curve of halo stars for the NFW halo profile	68
3.7	Plots with ringlike structure	72
3.8	Gas flaring with ringlike substructure	73
4.1	Schematic figure of the EGRET instrument	76

4.2	Third EGRET catalog	77
4.3	Survival probability of DMCs	80
4.4	Energy spectrum of the diffuse Galactic gamma radiation	83
4.5	Longitudinal gamma ray distribution of the NFW profile	86
4.6	Longitudinal gamma ray distribution of the 240 profile	87
4.7	Results of the Single Profile density model	89
4.8	Fit results of all halo profile combinations	91
4.9	Longitudinal distribution of the diffuse Galactic gamma radiation produced with the NFW-240 halo profile combination for photon energies above 500 MeV	94
4.10	Results of the Double Profile density model	95
4.11	Galactic rotation curve of the NFW-240 profile combination	97
4.12	Fermi/LAT apparatus	100
4.13	Differential spectra of primary cosmic rays and the diffuse Galactic gamma radiation	101
4.14	Photon energy spectrum measured with EGRET and Fermi/LAT	102
4.15	Gamma ray skymaps of the diffuse gamma radiation model from Fermi/LAT	103
4.16	Energy spectrum of the diffuse gamma radiation measured with Fermi/LAT	105
4.17	Longitudinal gamma ray distribution measured with Fermi/LAT	107
4.18	Schematic picture of the AMS-02 detector	108
4.19	Particle identification signals in the AMS-02 detector	109
C.1	Fit results for the NFW halo profile	132
C.2	Fit results for the BE halo profile	133
C.3	Fit results for the Moore halo profile	134
C.4	Fit results for the PISO halo profile	135
C.5	Fit results for the 240 halo profile	136
D.1	Fit results for the NFW-NFW profile combination	139
D.2	Fit results for the NFW-BE profile combination	140
D.3	Fit results for the NFW-Moore profile combination	141
D.4	Fit results for the NFW-PISO profile combination	142
D.5	Fit results for the NFW-240 profile combination	143
D.6	Fit results for the BE-NFW profile combination	144
D.7	Fit results for the BE-BE profile combination	145
D.8	Fit results for the BE-Moore profile combination	146
D.9	Fit results for the BE-PISO profile combination	147
D.10	Fit results for the BE-240 profile combination	148
D.11	Fit results for the PISO-NFW profile combination	149
D.12	Fit results for the PISO-BE profile combination	150
D.13	Fit results for the PISO-Moore profile combination	151
D.14	Fit results for the PISO-PISO profile combination	152

D.15 Fit results for the PISO-240 profile combination 153

List of Tables

2.1	Summary of the cosmological parameters	19
3.1	Luminous surface density contributions	57
3.2	Radial dependence of the different halo profiles	62
3.3	Parameters and experimental constraint for the density model	65
3.4	Local DM density results	66
3.5	Ring parameters	70
4.1	Definition of the spectral fit regions	81
4.2	Best parameters of the SP density model	85
4.3	Total Galactic mass, total matter density and total surface density at the Sun obtained from the SP density model	90
4.4	Best parameters of the DP density model	93
4.5	Total Galactic mass, total matter density and total surface density at the Sun obtained from the DP density model	96
4.6	Survival probability of DM clumps	98
4.7	Different DM distributions for the analysis of the Fermi/LAT data	106
B.1	Rotation velocities from publication by Sofue	120
B.2	Rotation velocities at the inner Galaxy	121
B.3	Rotation velocities within the Solar circle	122
B.4	Rotation velocities in the radial range from 9 to 15 kpc	123
B.5	Rotation velocities within the Solar circle	124
B.6	Rotation velocities from 8 to 17 kpc	125
B.7	Rotation velocities from 2 to 15 kpc	126
B.8	Averaged rotation velocities	126
B.9	Averaged rotation velocities	127
B.10	Averaged rotation velocities	127

B.11	Averaged rotation velocities	128
B.12	Averaged rotation velocities	128
B.13	Averaged rotation velocities	129
B.14	Averaged rotation velocities	129
D.1	Parameters of the remaining profile combinations	138

List of Abbreviations

BE	Binney-Evans (profile)
BHB	Blue-horizontal Branch
CGRO	Compton Gamma Ray Observatory
CMBR	Cosmic microwave background radiation
CDM	Cold Dark Matter
CR	Cosmic radiation
DM	Dark Matter
DMC	Dark Matter clump
EGBR	Extragalactic background radiation
EGRET	Energetic Gamma Ray Experiment Telescope
FSSC	Fermi Science Support Center
GC	Galactic centre
HDM	Hot Dark Matter
HI	Atomic hydrogen
HII	Ionized hydrogen
HWHM	Half-Width-Half-Maximum
LG	Local group
LSP	Lightest supersymmetric particle
MSSM	Minimal supersymmetric standard model
mSUGRA	Minimal Supergravity
MW	Milky Way
NFW	Navarro-Frenk-White (profile)
PISO	Pseudo-isothermal (profile)
RC	Rotation curve
SDSS	Sloan Digital Sky Survey
SM	Standard model of particle physics
SUSY	Supersymmetry
WDM	Warm Dark Matter

1
Introduction

In 1933 the Swiss astronomer Fritz Zwicky observed the COMA galaxy cluster and discovered that the amount of visible matter of the cluster galaxies is insufficient to explain their rotation velocities [1]. For that reason he introduced the concept of Dark Matter (DM) which only gravitationally interacts without emitting radiation. Later, observations of the rotation speed of gas and stars in spiral galaxies revealed more or less flat for practically all observed galaxies [2]. The DM content of these galaxies must be more widely distributed than the visible matter, since their velocity distributions do not show a Keplerian decrease, as expected from the visible matter in the centre.
The total matter density fraction of DM in the Universe is obtained from observations of the Cosmic Microwave Background Radiation (CMBR) with the Wilkinson Microwave Absorption Probe (WMAP) in combination with distance determinations from Type Ia supernovae (SNe) and baryon acoustic oscillations (BAO) to be about 23% [3]. The total matter density fraction of the baryonic matter is obtained to be about 4% which shows that the DM contribution can only be explained by non-baryonic matter. However, the nature of DM remains unknown until the present day.
A possible explanation for DM is a weakly interacting massive particle which is generally called WIMP. Obviously, the constituents of DM have to be massive due to their ability to interact gravitationally. A hint for the weak interaction of the DM particles comes from their spatial distribution. Since DM is distributed over large distances its energy losses due to interaction with other particles must be small. Otherwise, dense clusters of DM in the centre of a galaxy, like in case of the visible matter, would have been formed. Such a weakly interacting particle would be created in thermal equilibrium with all other particles in the early Universe. In this state its annihilation and production rate would be equal because of the high temperature of the early Universe. The temperature of the Universe dropped because of its expansion and at a certain temperature the expansion rate becomes higher than

the annihilation rate. In this case freeze-out occurs, i.e. the annihilation stops. The time when this happens is determined by the Hubble constant and the annihilation cross section. The remaining relic density at this time, called relic density, is inverse proportional to the annihilation cross section. From the Hubble constant and the relic density one finds indeed that the annihilation cross section is of the order of a weak interaction cross section [4–8]. The DM in galactic haloes is assumed to be a mix of a smoothly distributed DM component, which describes the DM distribution of individual WIMPs, and a clumpy DM component characterizing the distribution of local DM overdensities, so-called DM clumps (DMCs) or DM subhaloes. Local DM overdensities are resulting from primordial density fluctuations in the early Universe. Due to the gravitational interaction between the DM particles DMCs were formed after the freeze-out of the WIMPs. Subsequently, the DM clumps grew by merging with smaller clumps. This process, called hierarchical clustering, results in giant galactic haloes surrounding the luminous part of galaxies [9, 10]. The small clumps can be destroyed by tidal forces in the gravitational potential of the forming galaxies, thus forming a diffuse component of DM. Since not all clumps are destroyed during this process the DM distribution in a galactic halo can be split into a diffuse and a clumpy component, as shown by recent numerical simulations of structure formation [11, 12]. Recent N-body simulations show that the radial dependence of the density distributions of the two DM components might be different [13]. The gamma ray flux from DMA is proportional to the number density of the WIMPs. Therefore, the gamma ray flux should be dominated by the clumpy DM component because of the increased density in the DM subhaloes.

Two kinds of measurement of the properties of the WIMP are possible. Either the interaction of a WIMP with detector material is examined (direct searches) [14] or the final states of the annihilation of WIMPs are considered (indirect searches) [15]. The present thesis addresses the indirect determination of the Galactic DM density distribution using recent astronomical observations of the MW and the diffuse Galactic gamma radiation. The analysis of the diffuse Galactic gamma radiation measured with the Energetic Gamma Ray Emission Telescope (EGRET) in [16] showed an excess of gamma rays with a spectral shape of DMA and a spatial distribution of the gamma ray fluxes which is consistent with a cored halo profile in combination with a large scale structure of two rings if only a diffuse DM component is considered. However, the resulting DM density distribution yielded a high surface density which is incompatible with astronomical observations [17]. In this thesis diffuse Galactic gamma radiation is reconsidered for a Galactic DM composition of a smoothly distributed component and a component of DMCs.

The thesis is structured as follows. The theoretical framework is examined in Chapter 2. The standard model of cosmology, which describes the current understanding of the Universe, is introduced and observations to confirm this model are discussed. The formation of large structures like the Galactic DM halo and the MW are described in more detail. Possible DM candidates are summarised, the DM relic density is described and the estimation of the gamma ray flux produced by the diffuse and the clumpy DM component is given. After introducing the theoretical framework the DM density profile of the MW is considered in more detail in Chapter 3. There the local DM density, which is important for direct DM searches, is determined from current astronomical observations. First the different Galactic matter contributions are parametrised and the astronomical observations used to constrain the DM density distribution are described. Despite the new astronomical data an improvement to the viable range for the local DM is not found, because of the strong correlations between the visible and DM distribution. Furthermore, the smooth DM haloes expected from N-body simulations were found not to describe the structure of the rotation velocities in the outer Galaxy, which are increasing with radius. This inconsistency is solved by the introduction of a DM substructure composed of two rings – one at the inner Galaxy and one at the outer Galaxy – which is likely to be produced by the infall of a dwarf galaxy in the Galactic gravitational potential. Although in this case the local DM density is increased by a factor of about 3 compared to the density of the halo such a DM density distribution is found to be consistent with recent astronomical data from the "magic" ring of stars, called the Monoceros ring [18, 19], and the structure in the gas flaring [20], which can only be explained by a ring of DM. Subsequently, the gamma ray flux from DMA is considered in Chapter 4. The diffuse Galactic gamma radiation measured with EGRET is used to constrain the DM density distribution. Two models for the spatial DM distribution are introduced. In the first model, called Single Profile (SP) model, the smooth and clumpy DM components are distributed according to the same density distribution. In the second one the density distributions of the two DM components are assumed to be different, called Double Profile (DP) model. The clumpy (diffuse) DM component has a core (cusp) in the centre in agreement with recent high-resolution N-body simulations [13]. The SP model leads to high local DM densities which are incompatible with astronomical information. In contrast astronomical observations are compatible with a DP model which provides lower local DM densities. Preliminary data from the Fermi satellite, the successor of the EGRET satellite, are consistent with the DP density model. Finally, in Chapter 5 the results of the analysis are summarised and perspectives for further studies are given.

2
Theoretical framework

> *"The innocent and light minded, who believe that astronomy can be studied by looking at the heavens without knowledge of mathematics, will return in the next life as birds."*
> Plato (Timaeus, 91d)

2.1 The Cosmological Principle

Cosmology describes the Universe which is defined as a large system which is not part of a subsystem. Like all fields in physics cosmology is splitted into theoretical predictions and observations. The question is whether the observable Universe can be used to verify or falsify theoretical models for the entire Universe or not. For instance, measurements of the CMBR [3] have shown that the observable Universe is highly homogeneous but it is possible that homogeneity beyond the horizon may not be the case. This cannot be proven because we are not able to measure it.

In order to make the Universe a physical object which can be described by fundamental physical laws we have to suppose the *cosmological principle* [7]:

> **The observed part of the Universe represents the Universe in its properties and structures.**
> **The Universe is homogeneous in the large.**

Today one can add to this principle that there is no preferred direction in the Universe implying that the Universe is also *isotropic* at the large. There is no velocity of the Universe itself because it is the largest system and all measured velocities are measured with respect to the

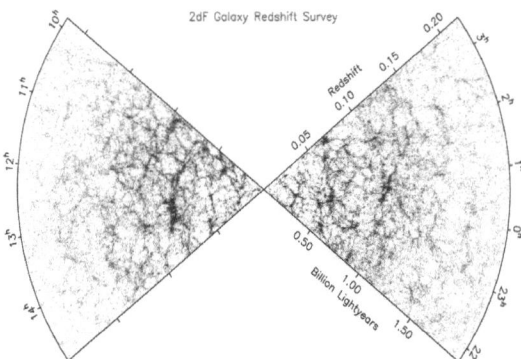

Figure 2.1: Map of the galaxy distribution produced from the complete 2dF Galaxy Redshift Survey. The map shows a homogeneous spatial distributions of the galaxies. Image taken from [22].

Universe. The cosmological principle allows to test theoretical conclusions by observations of the observable Universe. However, observations and measurements are only possible in the case of inhomogenities like stars, galaxies, galaxy clusters and so on. Measurements of the light emitted by galaxies show that these inhomogenities are uniformly distributed over large scales [21] as shown in Figure 2.1. Their structure is explained through homogeneous statistics and homogeneous laws of its genesis in Section 2.4.

In the eighteenth and nineteenth century the Universe was considered to be infinite, eternal and Euclidean. Measurements showed that stars are more or less at rest, with constant averaged luminosity per unit volume, and that there is no variability of seasonal and planetary phenomena. All this led to the assumption that the Universe is also static or stationary.

However, a static Universe filled with massive and radiative objects is in crucial conflict with simple cosmological considerations. Two famous problems of a static Universe will be explained in more detail in the following sections.

Newton's Paradox

One well-known conflict with a static Universe is Newton's paradox, in which the gravitational interaction of all objects in the Universe is considered. According to Newton's theory of gravity the gravitational force is given by

$$m\ddot{x} = -\text{grad } \phi = -4\pi G \text{ grad} \int_{R^3} \frac{\rho(x')}{\mid x - x' \mid} d^3x', \qquad (2.1)$$

where x represents the distance, ϕ is the gravitational potential and G is Newton's gravitational constant. The mass density distribution is given by $\rho(r)$. Three important properties of the force in Eq. (2.1) are that it is always attractive, it acts on all masses and its range is infinite. Although their strength decreases rapidly the forces between billions of galaxies are not negligible. Consequently, in a static and infinite old Universe all masses should have collapsed into a large, dense singularity which obviously did not happen. There are two explanations for the absence of the matter collapse. Either the age of the Universe is not infinite or some effect compensates the gravitational pull between massive objects. The motion of massive objects in a gravitationally bounded system prevents the gravitational collapse of the system. Good examples for this effect are the rotation of the Moon around the Earth, the rotation of the solar system around the Sun or the rotation of galaxies around the centre of mass in a galaxy cluster. Therefore, in order to solve Newton's paradox the Universe is either not static or its age is not infinite. In Section 2.2 it is shown that both cases are correct.

Olbers' Paradox

Another most commonly known conflict with a static Universe is named after the German astronomer Heinrich Olbers. He first asked why the night sky is dark, although this question was already mentioned by Marcellus Palingenius in 1570 and Edmond Halley in 1720. This question is justified since the range of electromagnetic radiation emitted by a star is infinite and in a static and infinite old Universe the light of an infinite number of stars would illuminate the night sky. Consequently, a static Universe would lead to a diverging averaged surface luminosity of the sky. A star which appears with a radius R in the night sky and which is the distance r away from Earth takes an area of about $\frac{R^2}{r^2}$ in the sky. The integration of a homogeneous and static Universe is given by

$$\Omega = \int_0^\infty 4\pi r^2 \frac{R^2}{r^2} dr \longrightarrow \infty. \qquad (2.2)$$

The solution of this problem is very difficult and not possible in a static Universe. The introduction of an expanding Universe only solved this problem since in this case just the light of the stars within the horizon of the Earth are visible in the night sky.

2.2 The Standard Model of Cosmology

In this section the introduction of a model for an expanding Universe called the standard model of cosmology and the evidence for this model are explained.
In 1915 Einstein published the theory of general relativity [23] introducing the famous field equations of gravitation

$$G_{\mu\nu} = 8\pi G\, T_{\mu\nu}, \qquad (2.3)$$

where $G_{\mu\nu}$ is the Einstein tensor representing the curvature of the four-dimensional spacetime and $T_{\mu\nu}$ is the energy-momentum tensor for all involved fields like matter, radiation, etc.

Einstein believed in a homogeneous, infinite and static Universe. However, his calculations showed that even in the simplest realisation of the Universe as a perfect fluid the field equations cannot be solved with the assumption of a static Universe. In 1917 Einstein showed that a physical solution for a three-dimensional sphere with constant positive curvature and density can be obtained if these equations[1]

$$G_{\mu\nu} + \Lambda\, g_{\mu\nu} = 8\pi G\, T_{\mu\nu}. \tag{2.4}$$

contained a constant Λ [24]. It is called cosmological constant and represents some kind of ground-value state of the scalar curvature. However, Einstein never solved his equations for a specific cosmological model, because of the tensorial character, which leaves too much freedom.

It was Friedmann who first solved Einstein's field equations by assuming that the Universe is homogeneous and isotropic [25]. This model, the Friedmann-Robertson-Walker (FRW) cosmological model, is called the standard model of cosmology (or the Big Bang model) and represents the current understanding of the evolution of the Universe. Friedmann showed that a positive curvature of space and a cosmological constant are not necessary for solving Newton's paradox when the distance scale is assumed to be explicitly time-dependent. In this scenario all distances vary with the same rate and all angles remain constant which let the Universe appear to be static. Hence a comoving volume can be defined where galaxies do not change their position according to their comoving coordinates. The expansion is thus homogeneous and a definition of an origin of the Universe is not possible. The variation of the distance scale is described by the scale parameter $R(t)$. Its evolution is given by the Einstein field equations

$$R_{\mu\nu} - \frac{1}{2}\mathcal{R} g_{\mu\nu} \equiv G_{\mu\nu} = 8\pi G T_{\mu\nu}. \tag{2.5}$$

These equations connect mass and energy (on the right side) to curvature of the four-dimensional space-time (on the left side) which is the fundamental concept of the theory of general relativity. In this expression $R_{\mu\nu}$ is the Ricci tensor or curvature tensor, $\mathcal{R}$ is the Ricci scalar, $G_{\mu\nu}$ is the Einstein tensor and $g_{\mu\nu}$ is the metric tensor which has in special relativity [26] the simple realisation diag(1,-1,-1,-1). The kinematics of the Universe are described by the Robertson-Walker metric

$$ds^2 = dt^2 - R(t) \cdot \left\{ \frac{dr^2}{1-kr^2} + r^2 d\theta^2 + r^2 sin^2\theta d\phi^2 \right\} \tag{2.6}$$

where r, θ and ϕ are the spatial coordinates of the comoving volume and k is the spatial curvature scalar which can be chosen to be $+1$, -1 or 0 for positive, negative or zero spatial curvature. The Robertson-Walker metric describes a homogeneous and isotropic Universe with constant curvature. In order to be consistent with the symmetries of the metric and

---
[1] $c = \hbar = k = 1$.

the cosmological principle the energy-momentum tensor has to be diagonal and the spatial components have to be equal. In the absence of matter in the energy-momentum tensor $T_{\mu\nu}$ is equal to zero while for an electromagnetic field $F_{\mu\nu}$ the form of $T_{\mu\nu}$ is given by

$$T_{\mu\nu} = \frac{1}{4\pi}\left(F_{\mu\lambda}F_{\nu\kappa}g^{\lambda\kappa} - \frac{1}{4}g_{\mu\nu}F_{\sigma\tau}F_{\kappa\lambda}g^{\sigma\kappa}g^{\lambda\tau}\right). \tag{2.7}$$

For a perfect fluid the components of the energy-momentum tensor are the time-dependent energy density $\rho(t)$ and pressure $p(t)$:

$$T^{\mu}_{\nu} = \text{diag}(\rho, -p, -p, -p) \tag{2.8}$$

The conservation of the energy-momentum tensor yields the first law of thermodynamics

$$d(\rho\, R^3) = -p\, dR^3 \tag{2.9}$$

which connects energy density and pressure in the comoving volume. The increase of energy in the comoving volume leads to the decrease of pressure in the volume. A simple equation of state $p = w\rho$, where w is independent of time, allows to make conclusions about different epochs of the Universe

$$\begin{aligned}
\text{radiation} \quad & (w = \tfrac{1}{3}) \quad \rightarrow \quad \rho \propto R^{-4} \\
\text{matter} \quad & (w = 0) \quad \rightarrow \quad \rho \propto R^{-3} \\
\text{vacuum energy} \quad & (w = -1) \quad \rightarrow \quad \rho \propto const.
\end{aligned}$$

In the early Universe the energy-momentum tensor was dominated by the radiation contribution. If there was an inflation epoch during the history of the evolution of the Universe the contribution of the vacuum energy was dominant. Today the Universe is dominated by the matter contribution.

The combination of Einstein's field equations and the Robertson-Walker metric yields the following equations

$$\frac{\dot{R}^2}{R^2} + \frac{k}{R^2} = \frac{8\pi G}{3}\rho \quad \text{and} \tag{2.10}$$

$$2\frac{\ddot{R}}{R} + \frac{\dot{R}^2}{R^2} + \frac{k}{R^2} = -8\pi G p \tag{2.11}$$

where Eq. (2.10) is the Friedmann equation. It does not contain a cosmological constant since Friedmann did not consider a static Universe. The difference of Eq. (2.10) and Eq. (2.11)

$$\frac{\ddot{R}}{R} = -\frac{4\pi G}{3}(\rho + 3p). \tag{2.12}$$

bears a first hint for a cosmological singularity, referred to as the Big Bang. At some time the scale factor $R(t)$ must have been zero if $\rho + 3p$ was always positive in the past.

An extrapolation beyond this singularity is not possible in the classical theory of general relativity.

The expansion rate $\frac{\dot{R}}{R}$ of the Universe is given by the Hubble constant H. With the Hubble constant the Friedmann equation can be rewritten

$$\frac{k}{H^2 R^2} = \frac{\rho}{3H^2/8\pi G} - 1 = \frac{\rho}{\rho_c} - 1 \equiv \Omega - 1, \quad (2.13)$$

where ρ_c is the critical density of the Universe. This equation shows the direct relation between the sign of the spatial curvature and the sign of the dimensionless energy density of the Universe

$$k = +1 \longrightarrow \Omega > 1$$
$$k = 0 \longrightarrow \Omega = 1$$
$$k = -1 \longrightarrow \Omega < 1$$

Consequently Ω is larger than one for a closed Universe, less than one for an open Universe and equal to one for a flat Universe as depicted in Figure 2.2. For closed models with $\Omega > 1$ a physical radius of the three-dimensional sphere can be calculated with

$$R_{\text{curv}} = \frac{H^{-1}}{|\Omega - 1|^{1/2}}. \quad (2.14)$$

Friedmann's model of an expanding Universe does not need a cosmological constant. However, vacuum fluctuations contribute to the energy density and thus counteract the gravitational force. Subsequently, the cosmological constant was added to the field equations again. The Robertson-Walker metric in combination with Eq. (2.5) leads to the Einstein-Friedmann-Lemaître (EFL) equations

$$\frac{\ddot{R}}{R} = -\frac{4\pi G}{3}(\rho + 3p) + \frac{1}{3}\Lambda \quad (2.15)$$

$$\frac{\dot{R}^2}{R^2} + \frac{k}{R^2} = \frac{8\pi G}{3}\rho + \frac{1}{3}\Lambda \quad (2.16)$$

which are analogous to Eq. (2.10) and (2.11). With the knowledge of the relation between ρ and p the EFL equations describe the dynamics of the Universe. Actual measurements of the pressure of the Universe show a very small value ($p/\rho \leq 10^{-4}$) validing the assumption $p = 0$ and $t = t_0$ in order to describe the present Universe. Thus the EFL equations change to

$$\Omega_\Lambda = \frac{\Omega_{m,0}}{2} - q_0 \quad \text{and} \quad (2.17)$$

$$\frac{k}{R_0^2 H_0^2} = \Omega_{m,0} + \Omega_\Lambda - 1 \quad (2.18)$$

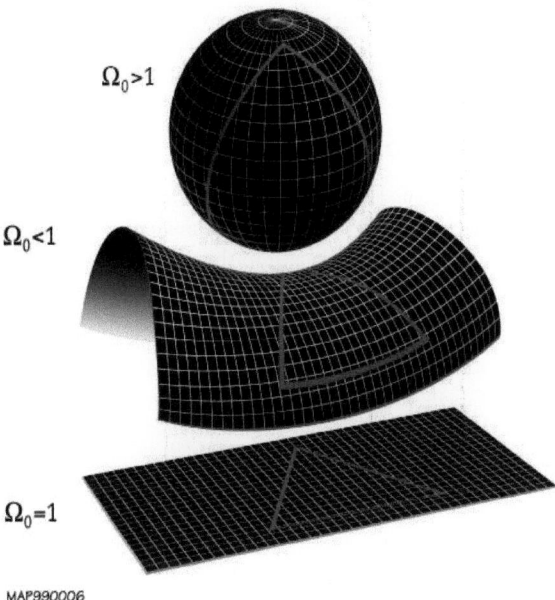

Figure 2.2: Visualisation of a closed Universe ($\Omega_0 > 1$) with positive spatial curvature, an open Universe ($\Omega_0 < 1$) with negative curvature and a flat Universe ($\Omega_0 = 1$) without curvature. Figure adapted from [27]

where Ω_Λ is the dimensionless vacuum energy density and q_0 is the deceleration parameter of the present Universe

$$q_0 = -\frac{\ddot{R}_0\, R_0}{\dot{R}_0^2}. \tag{2.19}$$

The hot Big Bang model is a very successful cosmological model. It is so robust that it is possible to make speculations about the Universe at times as early as 10^{-43} seconds after the Big Bang [6]. It is confirmed by several experimental measurements. The most important arguments are the measurement of the Hubble expansion, the measurement of the CMBR and the primordial nucleosynthesis. These arguments are also often referred to as "the three pillars" of the hot big bang model. In the following chapters the three pillars are considered in more detail.

2.2.1 Hubble Expansion

At about the time Einstein published his paper about a static and homogeneous Universe astronomers did the first measurements of redshifts of luminous objects. They defined the redshift z as the ratio of the detected wavelength to the emitted wavelength

$$1 + z = \frac{\lambda_{\text{obs}}}{\lambda_{\text{emit}}}. \tag{2.20}$$

A non-vanishing value of z was interpreted as the radial velocity of galaxies moving away from the Earth. In 1927 Edwin Hubble published results about the relation between the distance of galaxies and their radial velocity. He found that z is higher for more distant galaxies and that the data can be described by a simple linear equation

$$\frac{dr}{dt} = H_0 \cdot r \tag{2.21}$$

where r is the distance to the observed galaxy and gradient H_0 is the Hubble constant. Subscript 0 denotes the present value of a quantity.

From this simple relation one can make the following two conclusions. First, the Hubble expansion law fulfils the cosmological principle due to its translation invariant form

$$\frac{d(r - r')}{dt} = H_0(r - r').$$

Hence the expansion is homogeneous and a definition of an origin of the Universe is not possible. Using a time-dependent expansion parameter $R(t)$ the present expansion rate is given by the Hubble constant

$$H_0 = \frac{\dot{R}(t_0)}{R(t_0)}, \tag{2.22}$$

leading to

$$1 + z \equiv \frac{R(t_{\text{obs}})}{R(t_{\text{emit}})} \tag{2.23}$$

for the redshift. The second conclusion from the expansion law in Eq. (2.22) is that the current expansion rate of the Universe leads to a particular time t_0 in the past at which the distances between galaxies vanished. This is the first experimental hint for the Big Bang and (assuming that H_0 is independent of time) defines the age of the Universe. In the EFL model the age of the Universe can be estimated using Eq. (2.18) and the relation $\rho = \rho_0 \, (R/R_0)^3$. Then the age of the Universe is given by

$$t_0 = \frac{1}{H_0} \cdot \int_0^1 \frac{1}{\sqrt{1 - \Omega_{m,0} - \Omega_\Lambda + \Omega_{m,0}/x + \Omega_\Lambda x^2}} dx. \qquad (2.24)$$

This equation is valid as long as the Universe is dominated by the matter contribution ($p = 0$). The easiest estimation can be done in the Einstein-de Sitter model ($\Omega_\Lambda = 0$ and $\Omega_{m,0} = 1$) in which the age of the Universe is

$$t_0 = \frac{2}{3} \cdot \frac{1}{H_0}. \qquad (2.25)$$

These two approximations show that the order of the age of the Universe is roughly given by the Hubble time which is simply the reciprocal of the expansion rate. However, for different models of history of the expansion the values for the age of the Universe differ. Today the Hubble constant is given by

$$H_0 = 100 \, h \, \frac{\text{km}}{\text{s Mpc}} \qquad (2.26)$$

where the uncertainties of H_0 are hidden in little h. The current value for h from WMAP measurements [3] and thus the estimation of the age of the Universe is

$$h = 0.71 \pm 0.04$$

and

$$t_0 = (13.73 \pm 0.12) \cdot 10^9 \text{ yrs}$$

2.2.2 Cosmic Microwave Background Radiation

The standard model of cosmology is characterised by its high grade of isotropy and homogenity. The best evidence for these attributes of the Universe is the uniformity of the temperature of the cosmic microwave background radiation. In 1965 Arno Pensias and Robert Wilson first observed the CMBR when they tried to remove an excess of noise at a microwave antenna. The observed spectrum of the CMBR is a perfect black body spectrum since photons and charged particles were in thermodynamical equilibrium at the time of decoupling. Therefore its intensity follows the Planck radiation law with a maximal intensity at a wavelength of about 1 millimeter. The redshift of the photons ($z \approx 1000$) can be estimated

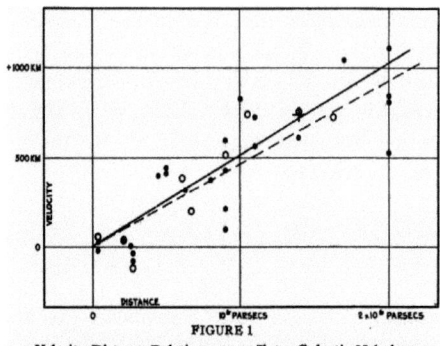

Figure 2.3: Hubble's velocity-distance relation. A linear correlation between velocity and distance is shown. The slope of the lines gives the Hubble constant H_0. The two lines represent different ways of correcting the motion of the Sun. Figure taken from [28].

with the ionisation energy of atomic hydrogen which is around 3000 K leading to an actual temperature of the CMBR of 2.73 K. Aside from the anisotropy caused by the motion of the Earth this temperature is highly isotropic. This high isotropy leads to a difficult problem, known as horizon problem, for the description of the formation of large scale structures in the Universe within the framework of the standard model of cosmology. This problem was solved with the introduction of a inflationary Universe (see Section 2.3).

However, the isotropy of the Universe is not perfect. Since 1992 measurements of the satellite experiments COBE [29] and WMAP [3] allow the mapping of the anisotropy of the CMBR. A sky map of the temperature fluctuations from WMAP is shown in Figure 2.4. The fluctuations of the temperature correspond to fluctuations of density at the time of decoupling. The temperature is higher in regions with higher density. There the photons decouple later from the thermodynamical equilibrium and therefore have a smaller redshift z. The temperature fluctuations can be expanded in spherical harmonics:

$$\frac{\Delta T}{T} = \sum_{l=2}^{\infty} \sum_{m=-l}^{l} a_{lm} Y_{lm}(\theta, \phi) \qquad (2.27)$$

Due to the motion of the Earth the dipole anisotropy of the temperature fluctuations is very large and has to be subtracted. The maximal amplitude of the quadrupole component is about 30 μK which is approximately 10^{-5} of the averaged temperature. The angular average of the temperature fluctuations is defined as

$$C_l = \langle a_{lm} a_{lm}^* \rangle = \frac{1}{2l+1} \sum_{m=-l}^{l} a_{lm} a_{lm}^* \qquad (2.28)$$

Figure 2.4: Map of the temperature anisotropies in the CMBR measured with WMAP [30]. The difference between black and red spots is approximately 10^{-5} of the averaged temperature of 2.73 K.

The coefficients C_l are shown as a function of the multipole moment l in Figure 2.5. This power spectrum of the CMBR anisotropies is characterised by several acoustic peaks which correspond to the compression and rarefunction of the baryon fluid. The odd numbered peaks are sensitive to the compression and therefore to the total energy density of the Universe. The even numbered peaks correspond to the maximal rarefunction of the baryonic matter which explains how the far plasma rebounds under radiation pressure. The physical reason for these peaks are acoustic waves which result from the competing actions of the gravitational attraction and the interaction pressure in the plasma at the time of decoupling. Therefore, the CMBR is a snapshot of the oscillating photon radiation field at the time of decoupling. Assuming a model for the origin of the anisotropies the shape of the power spectrum and the relative position and relative heights of the acoustic peaks can be used to constrain the values of cosmological parameters.

The power spectrum is relatively flat for $l < 20$ which shows that the temperature fluctuations are independent of the angular scale for angular differences larger than 10 degrees. The position of the first peak at $l \approx 200$ corresponds to an angular difference of approximately one degree which is the maximal distance between space regions which are in causal connection at the beginning of the matter dominated phase. This is in good agreement with a flat Universe ($\Omega = 1$). For a closed Universe ($\Omega > 1$) the first peak would be at larger angles while an open Universe ($\Omega < 1$) would be characterised by a first peak at smaller angles. The height of the first peak relative to the plateau at small values of l is sensitive to the total energy density of the Universe. This constraints the matter content Ω_m of the Universe. The height of the first peak is larger for smaller values of Ω_m since this implies

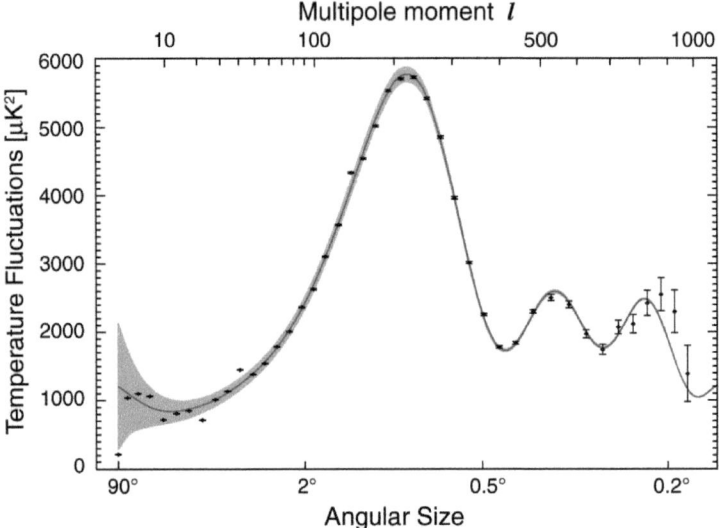

Figure 2.5: The angular power spectrum of the CMBR brightness fluctuations measured with WMAP is shown. The angular size of the fluctuations is given on the x-coordinate at the bottom while the multipole moment is given at the top. The power spectrum is flat for $l < 20$. The first, dominant peak is located at $l \approx 200$ which agrees with a flat Universe. The baryonic matter content Ω_b corresponds to the ratio of the ratio of the height of the first to the height of the second peak. Figure taken from [31].

the radiation to mass ratio is larger at the decoupling. In this case the radiation pressure is higher compared to the gravitational attraction which results in more modes for the acoustic oscillations. The ratio of the heights of the first and the second peak determines the content of baryonic matter Ω_b in the Universe. The value of the ratio is larger for a larger content of baryonic matter since in this case the height of the second peak relative to the first peak is smaller. The reason for this effect is that a larger Ω_b leads to a larger maximal compression during the gravitational collapse which increases the height of the odd numbered peaks as well.

The analysis of the power spectrum of the CMBR shows the relative correlations between the different matter contributions to the totel energy density of the Universe. In order to obtain the total values of the baryonic density, the DM density and the density of the vacuum energy the WMAP observations have to be combined with other combinations. In [3] the 5-year WMAP results were combined with distance measurements from Type Ia supernove

(SNe) [32] and baryon acoustic oscillations (BAO) [33], which are both sensitive to the expansion of the Universe. In the distance measurements with SNe a supernova is treated as a standard candle with known luminosity which makes a comparison of the distance from the Sun to the SN with expections using the Hubble constant H_0 possible. If the SN is further away than expected an additional acceleration of the Universe by dark energy occured. In case of the BAO distance correlations in the distribution of galaxies are examined. In order to explain this correlation a point-like density perturbation in the early Universe consisting of DM, gas, photons and neutrinos has to be imagined. The neutrinos immediately began to stream out of the density perturbation because of their high velocity and their low interaction rate with other particles. Contrary to this, the DM stayed in the centre of the perturbation since it is only interacting gravitationally and it has a low intrinsic velocity. The gas and the photons were coupled to each other because the ionised gas formed a plasma and the propagation of the photons was prevented by scattering processes with electrons. This led to a spherical sound wave in the plasma with a sound velocity of about 57% of the speed of light. At the photon decoupling epoch the photons became free and the plasma began to combine into neutral atoms. Then the sound velocity dropped and the pressure wave slowed down. At the time when all photons leaked out of the gas perturbation the pressure wave stalled and an acoustic peak at a diameter of about 150 Mpc had been formed. The distance of the acoustic peak from the centre of the perturbation only depends on the speed of sound in the plasma. The ratio of the current distance of the acoustic peak to the distance of 150 Mpc at the time of the decoupling of the photons represents the current scale parameter.

In Figure 2.6 the correlation between Ω_Λ and Ω_m obtained from the 5-year results of the measurement of the CMBR with WMAP and the distance measurements from SNe and BAO are shown. The resulting cosmological parameters are given in Table 2.1.

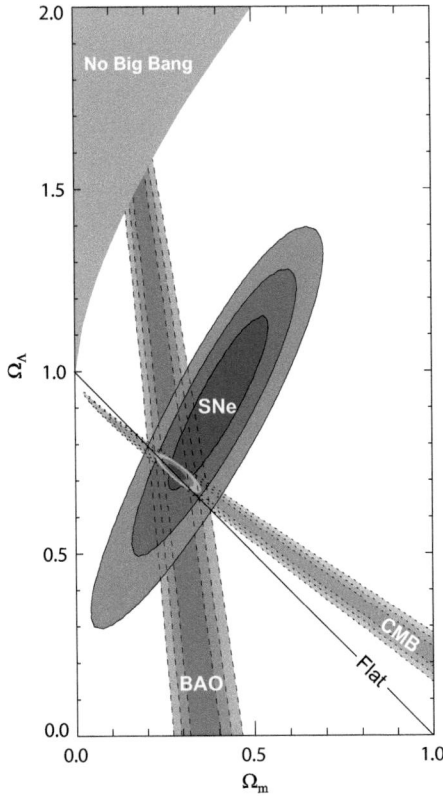

Figure 2.6: The contours of the 68%, 95% and 99.7% confidence level on Ω_Λ and Ω_m from the observation of the CMBR, from distance measurements with Type Ia supernovae (SNe) and baryon acoustic oscillations (BAO) are presented. The parameter w is assumed to be -1. Figure taken from [32].

Parameter	Symbol	Value	+ uncertainty	- uncertainty
Total density	Ω_{tot}	1.0052	0.0064	0,0064
Equation of state	w	-0.972	0.061	0.060
Matter density	$\Omega_m\,h^2$	0.136	0.0037	0.0036
Dark matter density	Ω_χ	0.233	0.013	0.013
Physical dark matter density	$\Omega_\chi h^2$	0.1143	0.034	0.0034
Baryon density	Ω_b	0.0462	0.0015	0.0015
Physical baryon density	$\Omega_b\,h^2$	0.02265	0.00059	0.0059
Light neutrino density	$\Omega_\nu\,h^2$	< 0.0065	(95 % CL)	
Hubble constant	H_0	70.1	1.3	1.3
Age of the Universe (Gyr)	t_0	13.73	0.12	0.12
Age of decoupling (yr)	t_{dec}	375938	3148	3115
Age of reionisation (Myr)	t_r	432	90	67
Redshift of matter-radiation equality	z_{eq}	3280	88	89
Redshift of decoupling	z_{dec}	1091.00	0.72	0.73
Redshift of reionisation	z_{ion}	10.8	1.4	1.4

Table 2.1: Summary of the cosmological parameters estimated from WMAP data combined with BAO and SNe observations [3].

2.2.3 Primordial Nucleosynthesis

The third experimentally verified prediction of the hot Big Bang model are the abundances of the light elements in the Universe. In 1946 Gamow introduced the idea of primordial nucleosynthesis [34]. Shortly before the CMBR was discovered the first estimation about the abundance of Helium (^{4}He) which was synthesised by the Big Bang was published by Hoyle and Taylor [35]. Unfortunately the complete calculation of the primordial abundances is a non-linear problem of a reaction network and can only be solved numerically. In the late sixties and early seventies of the twentieth century the first detailed codes to calculate this reaction network became available. In 1973 Wagoner wrote the so-called standard code for primordial nucleosynthesis [36]. Its numerical accuracy is better than 1% which is a great success for the model. A very important input parameter for the calculations is the ratio of neutrons to protons since nearly all neutrons in the Universe were used to form ^{4}He, the most tightly bound light nuclear state. Therefore, a simple picture can be used in order to estimate the abundance of ^{4}He. At very early times the Universe consisted of a quark gluon plasma where the quarks and gluons were in thermodynamical equilibrium. As the strong interaction rates became smaller than the expansion rate of the Universe the quarks froze out of this equilibrium and formed protons and neutrons. These particles were again in thermodynamical equilibrium maintained by the weak interactions

$$\begin{aligned} n &\longleftrightarrow p + e^- + \bar{\nu}_e, \\ \nu_e + n &\longleftrightarrow p + e^-, \\ e^+ + n &\longleftrightarrow p + \bar{\nu}_e. \end{aligned} \quad (2.29)$$

At this epoch ($t = 10^{-2}$ sec, $T = 10$ MeV) the energy density of the Universe was dominated by the radiation contribution ($w = 1/3$). As in the radiation-dominated phase the scale factor is $R(t) \propto \sqrt{t}$ and by using the Hubble constant as well as the Friedmann equation it is possible to get the time dependence of the temperature of the Universe

$$\begin{aligned} H^2 = \frac{\dot{R}^2}{R^2} = \frac{1}{4t^2} &= \frac{8\pi G}{3}\rho = \frac{8\pi G}{3} g_{\text{eff}} \frac{\pi^2}{30} T^4 \\ t &= \left(\frac{16\pi^3 G g_{\text{eff}}}{45}\right)^{-1/2} \cdot \frac{1}{T^2} \end{aligned} \quad (2.30)$$

where g_{eff} is the number of the relativistic degrees of freedom. These can be calculated from $e^\pm$, γ and 3 neutrino generations leading to $g_{\text{eff}} = 10.75$. When the weak rates are much larger than the expansion rate H the reactions in Eq. (2.29) establish equilibrium and the ratio of neutrons to protons is roughly one.

Later ($t \simeq 1$ sec, $T \simeq 1$ MeV) the 3 neutrinos decouple from the plasma and the $e^\pm$ pairs annihilate which leads to an increased photon temperature. Approximately at this time the

2.2 The Standard Model of Cosmology

rates of the weak interaction processes in Eq. (2.29) freeze out and the neutron-proton-ratio is

$$\left(\frac{n}{p}\right)_{\text{freeze out}} = \exp\left(-\frac{\Delta m}{T}\right) \simeq \frac{1}{6}. \tag{2.31}$$

Somewhat later ($t = 1$ to 3 min, $T = 0.3$ to 0.1 MeV) the neutrons and protons can form ^{2}H since the break-up of ^{2}H by the heated photon bath becomes lower than the fusion rate. At this time the neutron-to-proton ratio has decreased to about $1/7$ by the decay of neutrons and the number of relativistic degrees of freedom is $g_{\text{eff}} = 3.36$ because of the annihilation of the $e^\pm$ pairs. The production of ^{4}He via the chain reactions

$$^2\text{H}(^2\text{H}, \text{n})^3\text{He}(^2\text{H}, \text{p})^2\text{H},$$

$$^2\text{H}(^2\text{H}, \text{p})^3\text{H}(^2\text{H}, \text{n})^4\text{He and}$$

$$^2\text{H}(^2\text{H}, \gamma)^4\text{He},$$

starts at this time. However, the photodisintegration of ^{2}H is still very effective at these temperatures which is the reason why not much ^{4}He is produced. When the abundances of ^{4}H, ^{3}He and ^{3}H are built up the available neutrons are quickly bound into ^{4}He. Assuming that all neutrons are bound in ^{4}He its mass fraction can be estimated as

$$X_4 \simeq \frac{4n_{He}}{n_{tot}} = \frac{4(n_n/2)}{n_n + n_p} = \frac{2(n/p)}{1 + (n/p)}. \tag{2.32}$$

From the $n/p = 1/7$ ratio on finds $X_4 \approx 0.25$ in good agreement with measurements.
The predicted primordial abundances are a function of the baryon-to-photon ratio $\eta = n_B/n_\gamma$. The reason is that during the primordial nucleosynthesis the Universe was radiation dominated. The weak interaction rates in this epoch are proportional to the thermally averaged cross section $\langle \sigma|v|\rangle$, which is a function of temperature, and the number density of the concerning nuclear species $n_A(\eta, T) = (X_{A/A})\eta n_\gamma$. For larger values of η the abundances of the intermediate products ^{4}H, ^{3}He and ^{3}H build up earlier and the ^{4}He synthesis starts earlier as well. Assuming a larger neutron-to-proton ratio in this case more ^{4}He would be produced via nucleosynthesis. Therefore, in addition to the only free cosmological parameter η the primordial abundances are also sensitive to the physical parameters: the neutron half life $\tau_{1/2}(n)$ and the number of relativistic degrees of freedom g_{eff}. All weak interaction rates depend on the neutron half-life, $\Gamma \propto T^5/\tau_{1/2}(n)$, which is determined to be

$$\tau_{1/2}(n) = 10.5 \pm 0.2 \text{ min}.$$

An increase of $\tau_{1/2}(n)$ decreases all weak interaction rates that interconvert neutrons and protons and leads to a freeze-out of these particles at higher energies. According to Eq. (2.32) this leads to a different estimation for the abundance of ^{4}He. An increase of the second parameter g_{eff} leads to a faster expansion rate at the same temperature. In this case neutrons

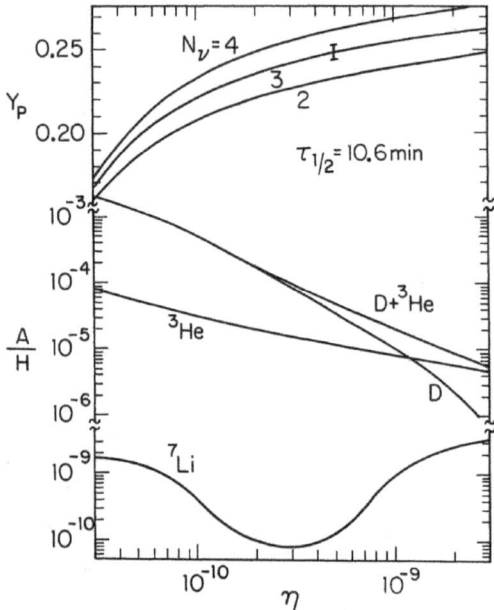

Figure 2.7: The predicted primordial abundances for the various light elements as a function of the baryon-to-photon ratio η. The primordial mass fraction for ^{4}He Y_P is shown for different values of g_{eff} and the error bar indicates the change of Y_P for $\Delta\tau_{1/2}(\text{n}) = \pm 0.2$ min. Figure was taken from [6].

and protons would freeze out of the thermodynamical equilibrium earlier, the neutron-to-proton ratio would be larger and more ^{4}He would be produced.

In Figure 2.7 the predicted primordial abundances is shown as a function of the baryon-to-photon ratio η. More details about primordial synthesis can be found in [6, 7].

2.3 Inflation

As mentioned above the standard model of cosmology is very successful and confirmed by observational measurements. However, there are still open questions which cannot be answered by the standard model. The observation of the CMBR, one of the most important verifications of the standard model, shows a very high smoothness of the Universe which,

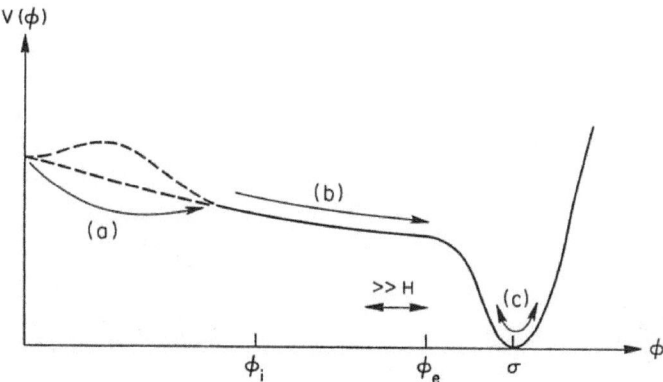

Figure 2.8: A schematic figure of the temperature potential $V_T(\phi)$. Three different parts are cleary visible. In part (a) a potential barrier is penetrated by quantum tunneling (if necessary). In part (b) the potential is flat enough to satisfy the slow-roll conditions. This is called the "slow-roll regime". In part (c) ϕ oscillates rapidly around the minimum. Taken from [6].

however, cannot be explained by the standard model. This problem is called the horizon problem. On the other hand on small scales the Universe is full of inhomogeneities like stars, galaxies, galaxy clusters, etc. The origin of these inhomogenities is not resolved in the standard model of cosmology. In 1981 Guth proposed an inflationary scenario in the very early Universe in order to solve these open questions [37]. Later in 1982 Linde [38] and Albrecht and Steinhardt [39] published an extended model called slow-rollover inflation, which represents our current understanding of the inflation of the Universe. The general idea of inflation is that at a certain time in the early Universe the vacuum energy was the dominant contribution to the energy density of the Universe. In this epoch small, smooth regions of a size smaller than $1/H$ can easily grow up to the size of the current observable Universe. The current model to describe inflation is based on spontaneous symmetry breaking (SSB) of a scalar field ϕ which depends on the temperature of the Universe. The phase transition of this symmetry breaking is characterised by an energy scale σ. If the temperature T is larger than σ the temperature potential $V_T(\phi)$ is minimal at $\langle\phi\rangle = 0$. When the Universe cools down and the temperature reaches the critical temperature T_c in first order phase transition a second minimum of the temperature potential at $\langle\phi\rangle = \sigma \neq 0$ is formed. Hence, the energy density of the Universe is dominated by the vacuum energy and the scale factor $R(t)$ increases exponentially. A schematic figure of such potentials is shown in Figure 2.8. For temperatures below the critical temperature the minimum at $\langle\phi\rangle \neq 0$ is the global minimum. After the negotiation of any barrier between $\langle\phi\rangle = 0$ and $\langle\phi\rangle = \sigma$ either by quantum tunneling or by the disappearance of the barrier at some temperature below T_c, the motion of the scalar field ϕ is described by

$$\ddot{\phi} + 3H\dot{\phi} + \Gamma_\phi \dot{\phi} + V'_T(\phi) = 0. \tag{2.33}$$

This equation is analogous to the differential equation for a ball rolling down a hill with friction. The course of the temperature potential shows that the motion of ϕ can be splitted into two parts. At first ϕ moves slowly along the potential before it is captured and oscillates around the new symmetry-breaking minimum. The first and the last term of Eq. (2.33) are well known from classical mechanics, the second term describes the red shifting of the momentum of ϕ and the third term descibes the oscillation of ϕ around $\langle\phi\rangle = \sigma$. Next these two different motions of the scalar field are considered in more detail.

In the first regime, the so-called slow-roll regime, $V_T(\phi)$ is quite flat and the kinetic energy of ϕ is much less than its potential energy. The motion of ϕ is dominated by the friction, $\ddot{\phi}$ is negligible and $\Gamma_\phi \dot{\phi}$ is not operative. Therefore, the equation of motion is reduced to

$$3H\dot{\phi} = -V'(\phi). \tag{2.34}$$

This means that the friction produced by the expansion of the Universe is equal to the acceleration produced by the slope of the potential. Therefore, in the slow-roll regime the motion of ϕ is uniform, which can only be obtained if a further acceleration of ϕ is negligible. In order to neglect $\ddot{\phi}$ the following conditions, the so-called slow-roll conditions, are required

$$|V''(\phi)| \ll 9H^2 \simeq 24\pi \frac{V(\phi)}{m_{\text{Pl}}^2},$$
$$\left|\frac{V'(\phi) m_{\text{Pl}}}{V(\phi)}\right| \ll \sqrt{48\pi}. \tag{2.35}$$

In the first condition the kinetic energy of ϕ needs to be much smaller than the potential energy, so that

$$H^2 \simeq \frac{8\pi}{3 m_{\text{Pl}}^2} V(\phi). \tag{2.36}$$

For any potential which is flat enough to satisfy these two conditions the scale factor R(t) grows very strongly until ϕ is captured and oscillates around the minimum.

In the second regime, the rapid oscillation regime, ϕ performs damped sinusoidal oscillations around $\langle\phi\rangle = \sigma$ with a frequency $\omega^2 = V''(\sigma) \gg H^2$. The equation of motion is given by

$$\ddot{\phi} + 3H\dot{\phi} + \Gamma_\phi \dot{\phi} = 0 \tag{2.37}$$

where the minimal potential energy is set to zero. These oscillations correspond to a condensate of zero-momentum ϕ particles of mass $m_\phi = V''(\sigma)$, which decay due to quantum particle creation into lighter particles that couple to ϕ. The damping of the oscillations which produces lighter particles leads to a reheating of the Universe called defrosting phase.

The decay width of the ϕ particles is represented by Γ_ϕ. Assuming that ϕ decays into two very light fermions which couple to ϕ the decay width is given by

$$\Gamma_\phi = \frac{h^2 m_\phi}{8\pi} \tag{2.38}$$

where the coupling strength is given by h. For $t \simeq \Gamma_\phi^{-1}$ the ϕ particles begin to decay rapidly which is the beginning of the radiation-dominated phase in the standard model of cosmology. For further reading about the open questions in the standard model of cosmology and inflation see [6, 7].

2.4 Structure formation

A closer look at the Universe shows that at small scales it is not homogenous at all. For instance, in galaxy clusters the density is 10^2 to 10^3 times higher than the averaged density in the Universe. In galaxies the density is even higher - approximately 10^5 times the averaged density of the Universe. On the contrary the isotropy in the CMBR and the galaxy distribution show that the Universe is smooth at large scales.

The question is why the Universe looks different for small and large scales. Cosmologists have a general picture for what happened during the evolution of the Universe in order to produce such a difference. In this scenario small, primeval density inhomogenities grew via gravitational instability into large inhomogenities like galaxies, galaxy clusters, superclusters and voids we observe today. These density fluctuations are defined as

$$\delta(\vec{x}) \equiv \frac{\delta\rho(\vec{x})}{\bar{\rho}} = \frac{\rho(\vec{x}) - \bar{\rho}}{\bar{\rho}}, \tag{2.39}$$

where $\rho(\vec{x})$ is the local density at the position $\vec{x}$ and $\bar{\rho}$ is the averaged density of the Universe. Since $\delta(\vec{x})$ is a scalar quantity it is possible to use comoving or physical coordinates. The Fourier expansion of the fluctuations is given by

$$\delta(\vec{x}) = \sum_{m,l,n=-\infty}^{\infty} exp(-i\vec{k}\cdot\vec{x})\, \delta_k \rightarrow \frac{V}{(2\pi)^3} \int_{\text{Vol}} \delta_k\, exp(-i\vec{k}\cdot\vec{x})\, d^3k, \tag{2.40}$$

$$\delta_k = V^{-1} \int_{\text{Vol}} \delta(\vec{x})\, exp(i\vec{k}\cdot\vec{x})\, d^3x, \tag{2.41}$$

where $V = L^3$ is the volume of the fundamental cube and $\vec{k}$ is the wavenumber for the modes enclosed in this cube. For values of L comparable to the length scales of the problem the spectrum of modes within this cube turns to a discrete spectrum with discrete wavenumbers for the space coordinates. The comoving wavelength of a density fluctuation is given by

$$\lambda = \frac{2\pi}{|\vec{k}|} = \frac{2\pi}{k}. \tag{2.42}$$

The physical values of the wavenumber and wavelength are $k_{\text{phys}} = k/R(t)$ and $\lambda = R(t) \cdot \lambda$. In the radiation-dominated phase of the Universe the density inhomogeneities are small as growing is inhibited by the radiation pressure of the photons. Once the Universe is dominated by the matter component the density fluctuations start to grow via the Jeans, or gravitational, instability (see Sect. 2.4.1) into large inhomogenties. So the time of matter-radiation equality is the initial epoch of structure formation. In order to find a suitable model for structure formation the initial conditions at that time have to be known. Therefore, it is important to know the total amount of non-relativistic matter in the Universe and its composition (cosmological constant, baryonic matter, Dark Matter, etc.). For a Universe dominated by Cold Dark Matter (CDM) consisting of WIMPs the DM decouples from radiation first while the baryons are still strongly coupled to the photons. So the density inhomogeneities in DM start to grow while the baryonic inhomogeneities cannot grow until the baryons decouple from the radiation. The structures formed by DM are small and dense because the DM particles are slow and massive which means that their gravitational energy is dissipated very fast. These small structures grow via hierarchical clustering to larger structures. This case is called bottom-up scenario. On the other hand, when DM consists of light, relativistic particles, so-called Hot Dark Matter (HDM), like neutrinos the baryonic matter decouples before the DM component. However, after the decoupling of the neutrinos they cannot dissipate their gravitional energy since they are very light, fast and weakly interacting. So they form larger and less dense structures than the CDM. Smaller structures are formed by the collapse of these large structures which is the reason why this is called top-down scenario. A further initial condition which has to be known is the spectrum of the primeval density perturbations. Usually it is expected that the spectrum is isotropic which means that it only depends on the wavenumber k. Since there is no definite model for the primeval fluctuations it is conventional to use a simple power law in order to parametrise the spectrum of primeval density fluctuations

$$|\delta_k|^2 = A \cdot V \cdot k^{n_p}, \qquad (2.43)$$

where n_p is the primeval power spectrum index. If the density fluctuations are Gaussian, as expected from inflationary models, any statistical quantity can be specified in terms of such a spectrum. In this case the density contrast is given by

$$\left(\frac{\partial \rho}{\rho}\right) = \frac{1}{\sqrt{V}} \cdot \frac{k^{3/2} \cdot |\delta_k|}{\sqrt{2\pi}} = A \cdot M^{-\alpha}. \qquad (2.44)$$

The mass of the density spectrum is characterised by M. The power index α can be calculated with $\alpha = 1/2 + n_p/6$. A special power spectrum with $\alpha = 1$ ($n_p = 3$) is the Harrison-Zel'dovich spectrum. This spectrum is predicted by inflationary models of the expansion of the Universe and describes density fluctuations with constant curvature. The type of the density fluctuations at the time of matter-radiation equality is also important for a structure formation model. It has to be distinguished between a curvature (or adiabatic) type which corresponds to fluctuations in the spatial curvature and an isocurvature type which corresponds to local variations in the equation of state. Here, the isocurvature type is explained in more detail. In this case pressure variations in the equation of state lead to density perturbations. Generally the growing of an overdense region is splitted into two parts

$$\frac{\delta\rho}{\rho} \propto \begin{cases} R & \delta\rho/\rho \lesssim 1 \text{ (linear regime)} \\ R^n(n \gtrsim 3) & \delta\rho/\rho \gtrsim 1 \text{ (non-linear regime)}. \end{cases}$$

In the linear regime the density distribution is still small and expands with the expansion rate of the Universe. When it enters the non-linear regime it separates from the expansion of the Universe and evolves like a small, separated and closed Universe with its own Friedmann equation. After expanding to the maximal radius it recollapses under formation of a gravitationally bound object. In an open Universe a small overdensity is insufficient to make such regions supercritical and they will expand eternally. On the other hand, a closed Universe with a total energy density bigger than unity will collapse before the overdensities become supercritical. Therefore a near-critical Universe is a condition for this simple explanation of structure formation in the Universe. The collapse of the overdensity is described by the spherical collapse model. There the overdensity shrinks by a factor of two during the collapse and increases its density by a factor of eight.

N-body simulation of the structure formation of the Universe [40, 41] show that such formation processes result in a filamentary cosmic DM distribution as shown in Figure 2.9.

2.4.1 Jeans instability

In this section a short overview about the Jeans (or gravitational) instability is given. In this model the expanding Universe is considered as a perfect, expanding fluid. The Newtonian motion of a perfect fluid is described by the equation of continuity, the Navier-Stokes equation and the Poisson equation

$$\begin{aligned} \frac{\partial \rho}{\partial t} + \vec{\nabla} \cdot (\rho \vec{v}) &= 0, \\ \frac{\partial \vec{v}}{\partial t} + (\vec{v} \cdot \vec{\nabla})\vec{v} + \frac{1}{\rho}\vec{\nabla}p + \vec{\nabla}\phi &= 0, \\ \nabla^2 \phi &= 4\pi G \rho. \end{aligned} \quad (2.45)$$

Here, ρ is the matter density, p the matter pressure, $\vec{v}$ the local fluid velocity and ϕ the gravitational potential. In an expanding Universe these values are given by

$$\begin{aligned} \rho &= \rho(t_0) \cdot R^{-3}(t) \\ \vec{v} &= \frac{\dot{R}}{R}\vec{r} \\ \vec{\nabla}\phi &= \frac{4\pi G \rho}{3}\vec{r}, \end{aligned} \quad (2.46)$$

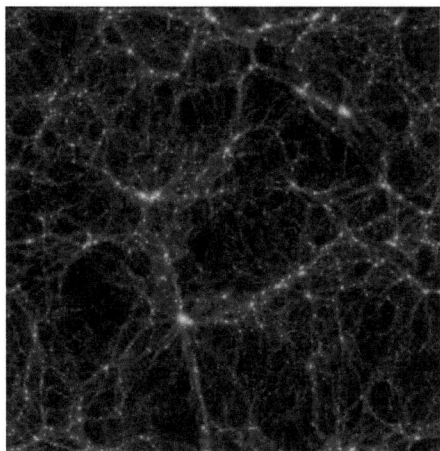

Figure 2.9: An example of the filamentary structure of the mass distribution of the Universe. Figure adapted from [42].

where $R(t)$ is the scale parameter of the usual Friedmann equation. Assuming that the perturbations of these values are adiabatic, i.e. that there are no spatial variation in the equation of state, in first order the perturbations satisfy Eq. (2.45). With the Fourier expansion of ρ, $\vec{v}$ and ϕ, shown in Eq. (2.40), the first-order equations of (2.45) become

$$\frac{d(R\vec{v}_\perp(\vec{k}))}{dt} = 0,$$

$$\vec{v}_\parallel(\vec{k}) = \frac{R(t)}{ik}\dot{\delta}_k + \frac{const}{R(t)},$$

$$\ddot{\delta}_k + 2\frac{\dot{R}}{R}\dot{\delta}_k + \left(\frac{v_s^2 k^2}{R^2} - 4\pi G\rho_0\right) = 0, \quad (2.47)$$

where $\vec{v}_\perp$ is the rotational and $\vec{v}_\parallel$ the irrotational ($\nabla \times \vec{v}_\parallel = 0$) component of the perturbated velocity field $\vec{v}(\vec{k})$. The unperturbated density of the fluid is characterised by ρ_0 and the sound velocity in the fluid is given by $v_s^2 = (\partial p/\partial \rho)_{\text{adiabatic}}$. These equations show that the rotational modes are independent of the matter perturbations. So, for structure formation only the irrotational modes are important. The Poisson equation in (2.47) turns out to be a wave equation and the Jeans wavenumber $k_J^2 = 4\pi G \rho_0 R^2/v_s^2$ separates gravitationally stable and unstable modes.

For $k \gg k_J$ the solutions are gravitationally stable and oscillate like a sound wave with decreasing amplitude. The interesting solutions for structure formation are the gravitationally unstable solutions with $k \ll k_J$ since these are the growing modes. On the assumption that the growth of density perturbations starts at the matter-radiation equality and that the Universe is spatially flat the Poisson equation has two independent solutions - a growing mode and a decaying mode. A density perturbation can be expressed as a superposition of these two solutions though the growing component becomes more important with increasing time.

The growth of density fluctuation is well described by Newton's gravitational theory as long as the modes are within the horizon of the Universe $\lambda_{\text{phys}} \ll H^{-1}$. For wavelengths of magnitude of H^{-1} or higher the description with Newton's theory is no longer valid and has to be replaced by Einstein's theory of general relativity. In this case both the Universe and the density fluctuations are described by similar Friedmann models with the same expansion rate. The Universe is assumed to be spatially flat ($k = 0$) whereas the density fluctuation is assumed to be spatially spherical with positive curvature ($k > 0$) and higher density. So the density perturbations are treated as a separated, closed Universe. Then the density contrast between these models is given by the curvature of the density perturbation

$$\delta = \frac{\rho' - \rho_0}{\rho_0} = \frac{k/R^2}{8\pi G \rho_0/3}, \qquad (2.48)$$

where the overdensity is characterised by ρ'. As discussed in Section 2.2 in the radiation dominated epoch the matter density ρ is proportional to R^{-4} which means that the density constrast is proportional to R^2. On the other hand, in the matter dominated phase the matter density is proportional to R^{-3} and thus the density contrast is proportional to R. Therefore, in the framework of general relativity density fluctuations grow with a time-dependence of $\delta \propto t$ for the radiation dominated epoch and $\delta \propto t^{2/3}$ in the matter dominated phase.

Further information about structure formation and the Jeans instability can be obtained from [6, 7, 43].

2.5 The Milky Way

Galaxies in the Universe have been observed since the development of the first telescopes. In those days they were called *nebulae* because they appeared as fuzzy objects on the night sky. So the early observed nebulae like the CRAB nebula are called after the star constellation where they appear. Since 1920 the existence of other galaxies has been established. In 1936 Edwin Hubble introduced a scheme in order to categorise galaxies in his book "The Realm of the Nebulae" [44]. According to this scheme galaxies can be sorted in four different categories: elliptical, lenticular, spiral and irregular. Elliptical galaxies are usually round, smooth and without any features. They can be very large with a radius of a few hundred kiloparsec[2] (kpc) and their luminosity can vary from 10^{-1} to 100 times the luminosity of

---

[2]Parsec (pc) is an abbreviation for parallax second. One parsec is the distance perpendicular to the ecliptic (rotation plane of the Earth around the Sun) for which the radius of the orbit of the Earth appears under one arcsecond. It holds 1 pc = 3.26 light years.

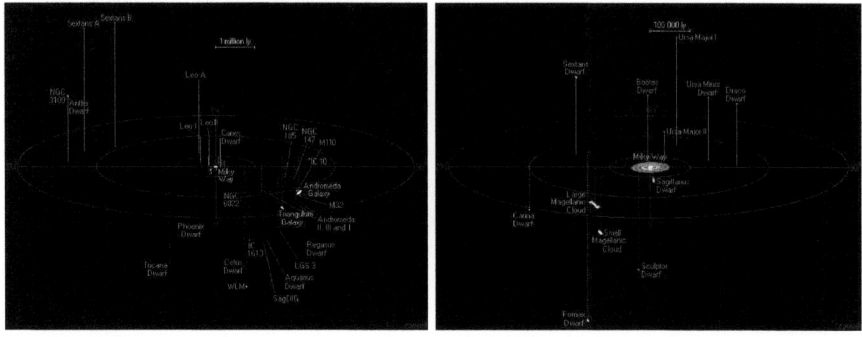

(a) Local Group   (b) Satellites of the Milky Way

Figure 2.10: A summary of the galaxies in the LG and the Galactic satellites is given. (a) The Local Group is composed of a large number of galaxies, even though its mass is dominated by the two largest constituents - the MW and the Andromeda galaxy. The Triangulum galaxy is the third largest galaxy in the LG with a mass of about 2% of the mass of the MW. (b) A summary of the dwarf galaxies orbiting the Galactic centre of the MW discovered until 1994 is shown. In this figure the Sagittarius dwarf galaxy, which was discovered in 1994, is the nearest Galactic satellite. This is slowing ripped appart by the MW. Figures taken from [45].

the MW. Lenticular galaxies consist of a rotating galactic disc and a rotating bulge but a spiral structure of the disc does not exist. Unlike to the discs of lenticular galaxies the galactic discs of spiral galaxies is structured with bright spiral arms which are outlined by clumps of bright stars. The spiral galaxies themselves can be further sorted by properties like luminosity of the galaxy or whether it has a central bar or not. The irregular categorie contains all galaxies which do not fit in the other categories. Today it is used only for small blue galaxies like dwarf galaxies without any organised structure.

The Milky Way (MW) is a typical spiral galaxy embedded in the local galaxy cluster, called Local Group (LG). The LG is an accumulation of galaxies within a distance of about 5 – 7 million light years, which are gravitationally bound to each other. About 95% of the visible matter in the LG is located in the MW and the Andromeda galaxy. The remaining galaxies are much smaller dwarf galaxies. A summary of the galaxies in the LG is shown in Figure 2.10a. The MW itself is surrounded by several dwarf galaxies as shown in 2.10b. The two nearest satellites are the Sagittarius dwarf [46] and the Canis Major dwarf [47] which both are ripped apart by tidal forces from the gravitational potential of the MW.

The Galactic bulge in the central region of the Galaxy is the gravitational centre of the Galaxy where gas density and the number density of stars is highest. The centre of the

Galaxy is believed to consist of a supermassive black hole of a mass of approximately 10^6 $M_\odot$ where $M_\odot = 1.99 \cdot 10^{30}$ kg. The most prominent feature of the MW is the thin and roughly circular Galactic disc which is visible as a luminous band on the night sky. The spiral structur consists of four large spiral arms: The Perseus arm, the Sagittarius-Carina arm, the Norma arm and the Crux-Scutum arm. The Sun is located at a Galactocentric distance of about 8.3 kpc in the Orion-Cygnus arm (local arm) which is a small lateral arm besides the Sagittarius-Carina arm. The spiral structure of the Galactic disc is summarised in Figure 2.11. The study of the stellar population of the outer disc [18] based on the observation of the fields at the Galactic anticentre with the Sloan Digital Sky Survey (SDSS) showed a ring structure outside the main spiral structure of the Galactic disc. This ring, which is unconnected to the spiral structure in the inner disc, is called outer ring or Monoceros ring. A possible origin of this structure could be the tidal disruption of a nearby orbiting Galactic satellite which is hidden within the Galactic disc. Further investigations on the ring structure showed it is localised to a Galactocentric distance of $\sim$ 15 to $\sim$ 20 kpc and has a scale height of about 750 pc [19]. However, the origin of the structure could not be ascertained. Shortly after that the remnants of a very nearby dwarf galaxy were discovered as an overdensity of M-giant stars in the Canis Major constellation [47]. The tidal stream of the disruption of the Canis Major dwarf galaxy is therefore believed to be a likely explanation of the Monoceros ring.

The radial dimension of the Galactic disc is approximately 20 kpc leading to a diameter of roughly 120,000 light years. Its density distribution drops exponentially with Galactocentric distance with a scale radius of 2 – 3 kpc. Furthermore, the disc is splitted into two parts - a thin and a thick disc. The star densities in both parts drop exponentially in vertical direction. The scale height for the thin disc is between 300 and 400 pc and for the thick disc between 1000 and 1500 pc. Nearly 95% of the disc stars and all of the young, massive stars are in the thin disc. The stars in the thick disc are older and poorer of heavy metals than the stars in the thin disc. The interstellar room between the stars is filled with interstellar gas and dust. These components lie within a thin layer of 100 pc from the midplane where the height scale increases with the galactocentric distance. The interstellar gas, however, is not a homogeneous medium. On small scales (smaller than about 1 kpc) it can be considered as a multiphase medium of a smooth gas component and a component consisting of gas clouds. These gas clouds host smaller gas clumps with even smaller subclumps. These hot and dense gas regions are the places where stars are born [48]. This multiphase gas distribution is analogous to the expected density distribution of the Dark Matter as will be discussed in Section 2.6. The composition of the interstellar gas consists of three components: ionised gas, neutral atoms and small molecules. Ionised gas consists of protons (in literature often referred to as ionised hydrogen (HII)), electrons and ionised "metals"[3] like oxygen, nydrogen and sulfur. Atomic hydrogen (HI), helium, carbon and oxygen are examples for neutral atoms and CO, HCN and CS for small molecules. All those components emit radiation either via

---

[3]In cosmology all elements except of hydrogen and helium are called "metals".

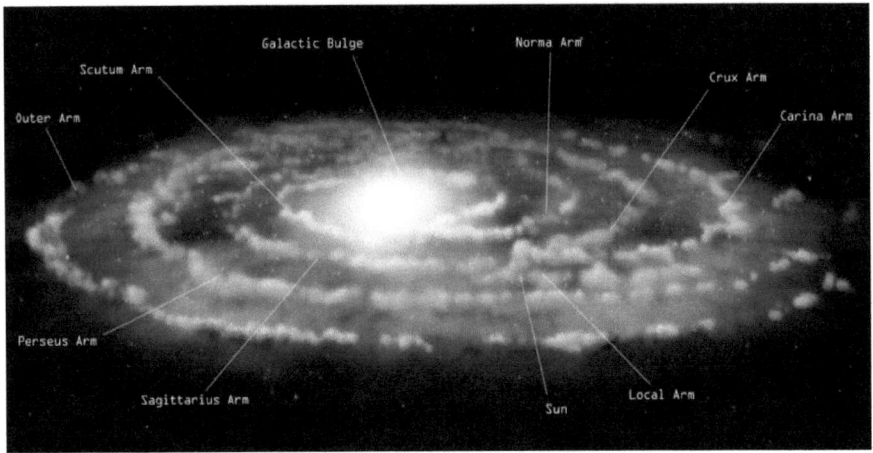

Figure 2.11: An illustration of the spiral structure of the Milky Way.

recombination, fine structure transition, hyperfine structure transition or vibration radiation. Measurements of the redshift of this radiation are used to determine the velocity distribution in the Galactic disc, called rotation curve of the MW.

2.5.1 Coordinate systems in the Milky Way

For the description of our Galaxy the spatial coordinates of Galactic objects have to be defined. In order to specify a special point in the galaxy three different coordinate system can be chosen: A spherical coordinate system centred at the Earth, a spherical system centred at the Sun or a cylindrical system centred at the GC of the MW. All three coordinate system will be shortly explained but only the last two systems were used in the analysis.

A spherical coordinate system centred at the Earth is quite easy to construct since it is just an extension of the terrestrial coordinate system with the latitudes δ and the longitudes α. The latitude can be measured with a telescope or a *Sextant* if the own latitudinal position is known. To measure the longitude the exact time of the measurement has to be known because of the rotation of the Earth. One degree in α corresponds to 15 minutes[4]. For the measurement of the distance parallaxe methods, photometric methods or spectrometic methods can be used. In such a coordinate system the rotation plane of the Sun around the Earth, called ecliptic, is $\delta = 23°27'$ inclined to the equatorial plane of the coordinate system.

---

[4]The solution of the longitude problem of the seafaring was solved in Greenwich, England, through the development of a clock which was possible to get over problems of a long journey on sea.

2.5 The Milky Way

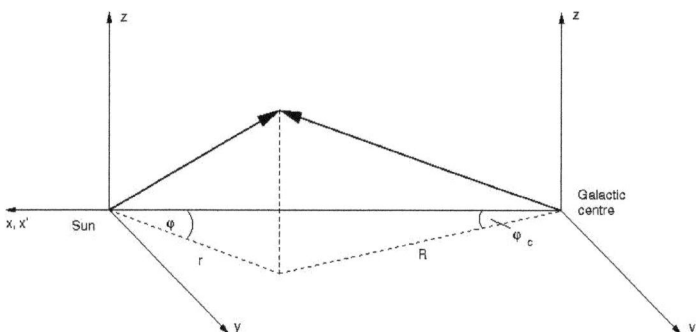

Figure 2.12: Transformation between the coordinate system centred at the Sun and system centred at the GC.

The Sun crosses this equator twice a year - on March 21st and September 23rd. These cross points are called equinox. Also the Galactic plane is inclined with respect to the equator of the coordinate system which is the reason why the disc of the MW is visible equally on the northern and southern sky. The GC is located at $\alpha = 17^h 42^m 24^s$ [5], $\delta = 27°24'$. These values are dependent on the year of observation since the rotation axis of the Earth changes slowly. Therefore, one has to take a reference for the coordinate system. Astronomers typically use the vernal equinox of the Sun of the year 1950, 2000 or the current year.

The spherical coordinate system centred around the Sun is quite similar to the system centred at the Earth. Here the longitudes are labelled with l and the latitudes are called b. The connection line between the Sun and the GC is defined as l = 0 and the Galactic plane defines b = 0. Longitudinal angles are in the range $-180° \leq l \leq 180°$ and latitudinal angles change between $-90° \leq b \leq 90°$. The longitudes increase counterclockwise and negative (positive) latitudes describe the region below (above) the Galactic disc. The region $0° < l < 180°$ is sometimes referred to as northern Galaxy because it is visible on the northern hemisphere of the Earth. Consequently, the region $-180° < l < 0°$ is called southern Galaxy.

The cylindrial coordinate system centred at the GC is based on the coordinates R, ϕ and z. The radius R is the Galactocentric distance of an object, z its height above the disc and ϕ its azimuthal angle with respect to the connection line between Sun and the GC.

The description of the position of a Galactic object with respect to the GC and the Sun is very important and necessary as density models of the Galaxy is always centred at the GC. However, oberservations of the properties of an Galactic object are always done from the position of the Earth. Figure 2.12 shows that the z and y coordinates are equal for

---

[5]For this representation the vernal equinox of the Sun in 1950 is taken as zero longitude. The angle is increasing in counterclockwise direction and 24 hours describe a complete circle.

both systems but the x coordinate is different. Within the Galactic plane the Galactocentric distance of a Galactic object can be calculated with the measured longitude ϕ, the distance to the Sun r and the distance R_0 between the Earth and the GC using the law of cosine. This transformation can be used to prove a chosen density distribution for the MW with experimental observations as will be described in Chapter 3. Further reading about Galactic coordinate systems can be done in [48].

2.6 Dark Matter

In the previous chapter we saw that the baryonic energy density is only about 4% of the total energy density of the Universe. The measurement of this value was obtained in the last quarter of the twentieth century. However, first hints on invisible matter were already found in the 1930s. In 1932 Jan Oort found that the density of matter near the Sun obtained from the kinematics of stars is larger than the density derived from the counting of stars [49]. This difference was explained with an additional amount of Dark Matter which was still not visible yet. One year later Fritz Zwicky observed the kinematics of the COMA galaxy cluster [1]. At this time measurements of the velocity distributions of the galactic disc, called rotation curve, became available for several galaxies. Zwicky used measurements of the rotation curves for seven galaxies in the COMA galaxy cluster and discovered that their rotation curves differ from the mean velocity of the cluster itself. With a crude approximation of the cluster radius he was able to calculate the total mass of the cluster. He estimated the total mass of the cluster obtained from the rotation velocities of the galaxies is about 400 times higher than the total mass derived from the observation of the luminous matter. Later in 1974 Ostriker and Einasto [50, 51] independently proposed that even isolated spiral galaxies are located in giant haloes which are several times larger than the radius of the luminous matter and contain most of the total mass of the galaxy. Eight decades after the establishment of DM in cosmological science its nature is still unknown. Candidates can be separated into baryonic and non-baryonic DM. Baryonic DM consists of small solid objects (dust), stellar objects which did not reach the lower mass limit for hydrogen burning of 0.08 $M_\odot$ (brown dwarfs) and stellar remnants like white dwarfs, neutron stars and black holes. However, the baryonic DM is still a part of the baryonic matter which is only a small part of the total matter of the Universe. So most of the DM is non-baryonic matter which can be further separated into HDM and CDM. The different candidates of non-baryonic matter are discussed in the following section.

2.6.1 Dark Matter Candidates

As its nature is completely unknown all possible explanations for DM need to be checked. The search for possible DM candidates is still an interesting field of science and all analyses which consider DM have to choose at least one of these candidates. Reviews on this issue are given in the publications [52–54].

Modified Newton Dynamics (MOND)

The concept of DM was introduced after the observation that kinematics of objects in gravitationally bounded systems like galaxy clusters or galaxies are not consistent with the kinematical predictions from the observation of the visible matter. Another way to explain this difference without introducing DM is the introduction of violations to Newton's law or the theory of general relativity [55, 56]. However, it turned out that it is difficult to find modifications which change gravity on all scales where the DM problem occurs. That is why this option of solving the DM problem is not satisfactory since the "standard" theory of gravitation describes a large amount of gravitational effects with enormous success.

Baryonic Dark Matter

The power spectrum of the CMBR is sensitive to the total energy density as well as to the baryonic matter content of the Universe. The evaluation of the measured spectrum shows that the baryonic matter content is much lower than the total matter content leading to the assumption that most of the matter in the Universe is non-baryonic. However, even the baryonic matter content is higher than the contribution of luminous baryons so baryonic DM also exists. These "hidden baryons" may be interstellar or intergalactic gases, which can be observed by the absorption of light from distant quasars, or massive and compact halo objects, so-called MACHOs, in the halo of the Galaxy. The observation of such objects is quite difficult. For this purpose a clever method is used where the intensity of the light of far away stars (located in the Large Magellanic Cloud or the Small Magellanic Cloud) is measured during the transit of a MACHO. Then the intensity of the star light rises because of the microlensing effect of the MACHO [57]. Unfortunately, it can not be detected which object has produced the lensing effect. Also very low-mass stars (M ≈ 0.1 $M_\odot$) can produce such lensing effects. In [58] it was shown that it is possible to identify events from stars down to a mass of about 10^{-5} solar masses.

Non-baryonic Dark Matter

Most of the DM density in the Universe is non-baryonic matter. Theoretical particle physics provides a large number of possible candidates for this matter contribution. Not all of these candidates are considered here but four of the most discussed particles will be explained.

Massive neutrinos
In contrast to all other particles provided by theoretical particle physics to be a DM candidate

the advantage of neutrinos is that they are known to exist. In the standard model of particle physics (SM) neutrinos are treated to be massless which is consistent with direct kinematical measurements. However, the observation of the solar neutrino flux by KAMIOKANDE [59], SNO [60], K2K [61] and KARMEN [62] gives evidence that neutrinos can oscillate in flavour. This effect does not seem to be solveable by modifications of the standard solar model [63] which indicates that neutrinos must have a non-zero mass. The best laboratory upper limit on the electron neutrino mass so far comes from tritium β-decay experiments [64]

$$m_{\nu_e} < 2.05 \text{ eV (@ 95\% C.L.)}. \tag{2.49}$$

Upcoming experiments like KATRIN [65] will reach a higher sensitivity on the electron neutrino mass so that a more precise value is expected within the next years. However, in combination with the small mass differences obtained from neutrino oscillation observations a sum of all neutrino masses of about 6 eV can be obtained [66]. The cosmological mass range allowed for this sum is more restrictive. From the observation of the CMBR with WMAP a sum of the neutrino masses is determined to be [3]

$$\sum_{i=1}^{3} m_{\nu,i} = 0.67 \text{eV}. \tag{2.50}$$

Massive neutrinos are predicted to have been thermally produced in the early Universe and decoupled at an energy of about 1 MeV. In this case their relic abundance depends on the sum of the different flavour masses

$$\Omega_\nu h^2 = \sum_{i=1}^{3} \frac{m_{\nu_i}}{93 \text{eV}}. \tag{2.51}$$

With the upper mass limit from Eq. (2.50) this leads to a relic neutrino density of $\Omega_\nu h^2 \lesssim 0.007$. This shows that the major part of the DM density in the Universe can not consist of massive neutrinos.

Axion
The axion is also often discussed as a DM candidate. In particle physics it was introduced in order to solve the CP violation of the strong interaction. When QCD was introduced as the fundamental gauge theory of the strong interaction it was found that the non-perturbative effects should induce a large CP violation in the strong interaction. It was Peccei and Quinn [67] who introduced an additional spontaneously broken, global symmetry in order to solve the strong CP violation problem. The Goldstone of this broken symmetry is the axion. The upper limit on its mass obtained from laboratory searches, stellar cooling and the dynamics of the supernova 1987 A is $m_a \lesssim 0.01$ eV [54]. This mass is very low but the coupling of the axion to ordinary matter particles is so weak that it was never in the thermal equilibrium and it behaves like CDM today. The calculation of its relic abundance is

uncertain because it depends on assumptions on the production mechanism. Nevertheless in the axion mass range between 10^{-5} and 10^{-2} eV the axion passes all observational constraints.

Neutralino

One of the most promising DM candidates is provided by supersymmetric extensions of the SM. In supersymmetry (SUSY) new particles at the energy scale of about 1 TeV are introduced in order to solve the drawbacks of the SM [68]. In the minimal supersymmetric standard model (MSSM) each SM particle is associated with a supersymmetric partner with different spin. The superpartners of SM fermions are bosons and vice versa. The nomenclature for the new particles is a prefix "s" for the superpartner of the SM fermions (*scalar fermion*; e.g. selectron, stau, etc.) and a suffix "ino" for the superpartner of the SM boson (e.g. photino, Wino, Zino, etc.). Furthermore, a new multiplicative quantum number, the R-parity, is introduced for the interaction between SM particles and SUSY particles. For SM particles the R-parity is equal to 1 while for SUSY particles it holds $R = -1$. Assuming R-parity conservation SUSY particles can only be produced and destroyed in pairs and the lightest supersymmetric particle (LSP) is stable. In the minimal Supergravity (mSUGRA) models, in which SUSY is broken via gravity mediation between the visible and the hidden sector, the neutralino is the LSP. It is a mix of the neutrally charged photino, Zino and the neutral Higgsinos. It is commonly assumed that the neutralino was produced in a large number and that it was in thermal equilibrium in the early Universe. During the expansion of the Universe its number density decreased until its interaction rate became smaller than the expansion rate of the Universe. Then the neutralino froze out of the thermal equilibrium and produced a relic density. The neutralino is a Majorana particle and is assumed to be non-relativistic at the time of freeze out. The annihilation products are fermion and gauge boson pairs. The attractivness of the neutralino as a DM candidate comes from several particle properties: It is electrically neutrally charged so that it does not absorb or emit light, it is stable so that it survived the long time from the freeze out until the present day, it couples to gauge bosons and its mass explains the measured relic density for a large SUSY-parameter range [68].

WIMPzilla

Assuming that the DM particle is a thermal relic of the early Universe the maximal mass of the DM candidate is about 340 TeV. This is called unitary bound and was predicted in 1989 by Griest and Kamionkowski as a consequence of the maximal thermal annihilation cross section [69]. The current unitary bound from measurements of the cosmic microwave background is 34 TeV. WIMPzillas are superheavy DM candidates with masses larger than 10^{10} $M_\odot$. Consequently these particles were not in thermal equilibrium during freeze-out and therefore their relic abundance does not dependent on their annihilation cross section.

The WIMPzillas can be produced in several ways. Among others it can be produced via gravitational production between the inflationary and the matter-dominated Universe [6] or via the oscillation of the inflation potential during the defrosting phase after the inflation. A more detailed description of these production scenarios can be found in [70].

Kaluza-Klein Dark Matter

In 1921 Kaluza tried to unify electromagnetism with gravity by the introduction of additional components to the metric tensor [71]. This marks the birth of the idea of additional dimensions which appear at high energies. These theories are called unified extra dimension (UED) theories. In most of these UED models the ordinary 3 + 1 space-time is named "brane" which is embedded in the expanded $3+\delta+1$ space-time called bulk. In the simplest UED model focussed on DM a flat extra spatial dimension is introduced. The SM particles can propagate in this extra dimension and therefore obtain an additional contribution to their kinetic energy. So, for every bulk field a set of Fourier expanded modes exist which is called Kaluza-Klein (KK) states. In the brane the KK states appear as a tower of states with masses increasing with the mode number. Each of these new states contains the same quantum numbers. The lightest Kaluza-Klein particle (LKP) is associated with the first KK excitation of the photon refered to as $B^{(1)}$. With a mass of the LKP between 400 and 1200 GeV the observed DM quantities can be explained. The results of the LKP calculations are sensitive to the spectrum of the first excitation of other particles. Coannihilation with the next to the lightest KK excitation (NLKP) is possible as well. So, the spectrum of the LKP is calculated to one-loop level so far. Unlike to the neutralino the LKP has bosonic character so there is no helicity suppression in its annihilation. So the annihilation into fermion pairs is more efficient than in the case of the neutralino. For further reading about extra dimensions and Kaluza-Klein particles see [72].

2.6.2 Relic Density

From the WMAP microwave background experiment, combined with other sets of data, that the relic density of DM in units of the critical density is $\Omega_\chi h^2 = 0.1143 \pm 0.0034$. A particle physics model for the describtion of DM has to fulfill this constraint. DM is likely consist of WIMPs. Their relic density was formed when these particles froze out of the thermodynamical equilibrium. In this epoch the time evolution of the WIMP number density n_χ can be approximated with the Boltzmann equation

$$\frac{dn_\chi}{dt} = -3Hn_\chi - \langle \sigma v \rangle \cdot (n_\chi^2 - n_{\chi,eq}^2), \tag{2.52}$$

---

[6] This production is similar to the generation of the density perturbations which are the starting point for the present large scale structures.

where H is the expansion rate of the Universe, $\langle \sigma v \rangle$ the thermally average annihilation cross section of the WIMP particle and $n_{\chi,\text{eq}}$ is the number density in the thermodynamical equilibrium. So the Boltzmann equation depends on the particle physics model which describes the interactions of the WIMP particles. Consequently, the problem of solving the Boltzmann equation is actually a problem of the determination of the annihilation cross section. However, using some simplified analytical considerations of the time evolution of the WIMP number density without specifing a model for the interactions the annihilation cross section of the WIMPs can be estimated. In thermal equilibrium the term proportional to the annihilation cross section cancels out and the time derivation of the number density is proportional to the expansion rate of the Universe $H = \dot{R}/R$ leading to $n_\chi \propto R^{-3}$. At early times when the temperature of the Universe was higher than the WIMP mass the scale radius $R \propto 1/T$ leads to $n_\chi \propto T^3$. Consequently the annihilation rate

$$\Gamma = n_\chi \cdot \langle \sigma v \rangle \qquad (2.53)$$

decreases with T^3 as well. When the annihilation rate Γ drops below the expansion rate of the Universe the WIMPs cease to annihilate. Then they fall out of the equilibrium and form a constant density, called relic density, which is given by

$$\Omega_\chi h^2 = \frac{m_\chi n_\chi}{\rho_c} \approx \left(\frac{3 \cdot 10^{-27} \text{cm}^3 \text{s}^{-1}}{\langle \sigma v \rangle} \right). \qquad (2.54)$$

Using WMAP measurements the cross section is roughly $3 \cdot 10^{-26}$ cm^3 s^{-1}. The numerical solution of the Boltzmann equation is quite difficult since many different annihilation channels (and eventually coannihilation processes) have to be taken into account for the calculation of the annihilation cross section. Two program codes which numerically calculate the relic density within the framework of supersymmetric particle physics are **Darksusy** [73] and **micrOMEGAs** [74]. A numerical solution of the Boltzmann equation is shown in Figure 2.13.

2.6.3 Annihilation of Dark Matter particles

The constituents of DM are still unknown. Nevertheless, theories of structure formation in the Universe give hints for the properties of a DM particle. In the framework of supersymmetrical particle physics the neutralino is a promising candidate for the WIMP. This particle is a Majorana particle with half-integer spin. In most of the supersymmetric particle models the neutralino is the lightest supersymmetric particle (LSP) which is stable if R-parity is conserved. A neutralino pair can annihilate into two SM particles. These annihilation products can be leptons, baryons or high energetic photons. The direct detection of electrons and protons is not possible since they immediately disappear in the sea of the cosmic radiation particles. However, the positrons, antiprotons and photons can be distinguished from the Galactic background. The Feynman diagrams of the main annihilation channels and their

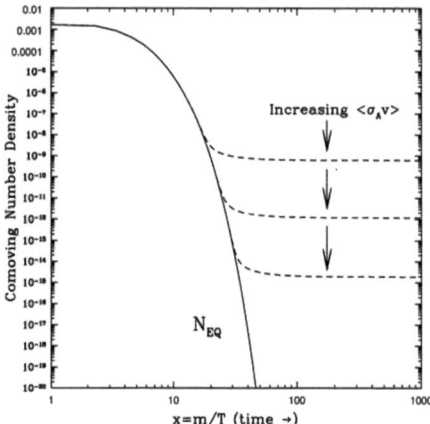

Figure 2.13: Numerical solution of the Boltzmann equation. Figure taken from [6].

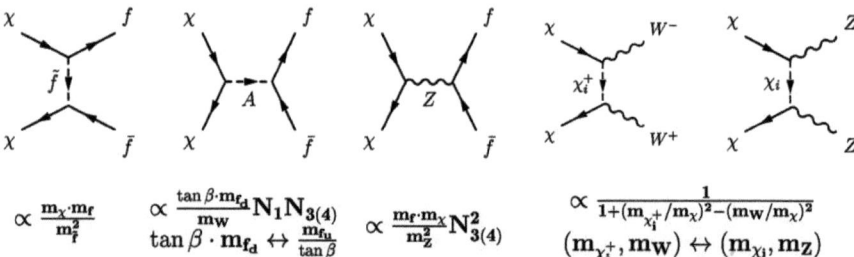

Figure 2.14: The main neutralino annihilation channels and their dependences on supersymmetric model parameters and masses are shown. The parameters N_{1-4} are the first row of the neutralino mixing matrix. Taken from [75].

dependence on the model parameters are shown in Figure 2.14. The Feynman diagrams show that the neutralinos are expected to annihilate into fermion-antifermion pairs and the gauge bosons of the weak interaction ($W^{\pm}$, Z) which produce a large amount of hadronic jets mainly consisting of neutral pions. The neutral pions almost always decay into two photons which is the reason why on average 30 to 40 photons are produced per WIMP annihilation process. On the assumption that the internal velocity of the WIMPs is very low, which is a good approximation for a WIMP mass of a few tens of GeV, the quarks in the final state of the annihilation are monoenergetic. In this case the energy spectrum of the gamma radiation resulting from the neutral pion decay only depends on the WIMP mass and the involved annihilation channels.

The CDM in the Universe is gravitationally unstable, i.e. it forms gravitationally bounded structures. This was shown in numerical simulations of the structure formation in the Universe [12, 40, 76], called N-body simulations, as well as in analytical calculations [10]. If the primordial density fluctuations at the time of the beginning of the matter dominated phase are large enough, bounded DM states at small scales, so-called DM clumps (DMC), can be formed. In this case the DM contribution of the MW consists of two different components - a smoothly and homogeneously distributed contribution, which will be hereafter referred to as "diffuse DM", and a clumpy distributed contribution called "clumpy DM". In the next two sections the annihilation fluxes of these contributions will be considered.

Diffuse Dark Matter

The diffuse DM component is given by the DM which is distributed in the halo of the MW. The DM particles in the halo are distributed according to a smooth density distribution called *halo profile*. For the DMA flux of the diffuse component ϕ_{diff} in a particular direction ψ with an energy E one has to integrate the squared DM number density along the line-of-sight (los)

$$\Phi_{\chi,diff}(E,\psi,\Delta\Omega) = \frac{\langle\sigma v\rangle}{4\pi} \cdot \sum_f \frac{dN_f}{dE} b_f \cdot \frac{1}{\Delta\Omega} \int_{\Delta\Omega} \int_{los} \frac{\langle\rho_\chi\rangle^2}{m_\chi^2} dl_\psi. \quad (2.55)$$

Here $\Delta\Omega$ is the solid angle in the direction ψ and dN_f/dE is the differential number of photons per annihilation for a particular final state at the energy E and b_f is the branching fraction of this final state. The integration has to be done over the product of the density distributions of the annihilating particles. This equation can be separated into a particle physics part depending on the properties of the DM particle and a cosmological part depending on the distribution of the DM. For the calculation of the cosmological part the density profile $\rho_\chi(r)$ is of crucial importance. An estimate of the density profile of the diffuse DM can be obtained by the comparison of the theoretical flux calculated with Eq. (2.55) and the experimentally measured flux of the diffuse Galactic gamma radiation.

Clumpy Dark Matter

Given the high clump density, most of the annihilation flux results from DM subhaloes. In case of CDM and a power index of $n_p \approx 1$ the small-scale clumps are the first gravitationally bounded states in the Universe with a minimal mass of $M_{\min} \sim 10^{-6}$ $M_\odot$ [10]. The larger clumps are formed later by hierarchial clustering. The small clumps are destroyed by tidal forces from the gravitational potential of the large clump. This effect changes the density profile and the mass of the large clump. However, this hierarchial clustering of small-scale clumps is a complicated process since the DM clumps are not totally virialised when they are captured by the next generation of larger clumps. This effect dominates the destruction of DMCs in the early stage of structure formation. During the clustering process it is also possible that small-scale clumps are not completely destroyed. Therefore, a large DMC hosts a few smaller clumps and its structure is similar to the structure of giant galactic haloes with masses of the order of 10^{12} $M_\odot$ which are also formed in this way. The baryonic matter follows the DM density distribution and forms stars and whole galaxies in the inner core of the DM halo where the DM density is highest. Detailed N-body simulations of galaxies of the size of the MW have shown that such a large halo can contain about 300,000 subhaloes. Due to the hierarchical clustering process up to four smaller generations of DMCs can be found in these subhaloes [11,12]. However, the number of subhaloes in the halo of the MW is much larger than the number of observed satellite galaxies of the Milky Way. This is known as the missing satellite problem. Up to the present day it is unclear why no galaxies are formed in most of the subhaloes of the MW. The problem can be solved either by assuming that the DM particles have a higher velocity dispersion, called Warm Dark Matter (WDM), which would suppress the production of small-scale structures and reduce the number of low mass subhaloes, or the formation of stars and galaxies is suppressed in the subhaloes by thermal feedback from the young galaxy, quasars or supernovae. The second explanation agrees with the large number of subhaloes but a better understanding of the prevention of star formation by thermal feedback is necessary in order to solve the problem.

For the calculation of the annihilation signal of a DMC the density profile of the clump is of crucial importance. Since the structures of a large clump and a galactic halo are similar according to N-body simulations, the easiest way is to assume a similar parametrisation of the density distributions. In both cases a power-law parametrisation is used. The parametrisation of the halo profile will be explained in Section 3.2.2. The annihilation flux of a DMC was analytically calculated in [10]. There the internal density profile of a clump was assumed as

$$\rho_{\text{int}}(r) = \frac{3-\beta}{3} \bar{\rho} \left(\frac{r}{R_{\text{cl}}}\right)^{-\beta}, \qquad (2.56)$$

where R_{cl} is the radius of the clump and r is the distance from the centre of the clump. The value of the power index β is the range of 1.7 to 1.9. A further value important for the description of a DMC is the core radius r_c. The core is the region with the highest density in the centre of the clump. According to [10] a theoretical estimation of the ratio of the core radius to the clump radius can be obtained from the density fluctuation δ_{eq} at the beginning of the matter-dominated phase. The ratio of the core radius and the clump radius is then given by the value $x_c = \frac{r_c}{R_{cl}} \approx \delta_{eq}^3$. Analogous to the diffuse DM component with the internal density profile the annihilation flux of the DMC is given by

$$\dot{N}_{cl} = 4\pi \int_0^\infty r^2 \frac{\rho_{int}^2(r)}{m_\chi^2} \langle \sigma v \rangle \, dr, \quad (2.57)$$

$$= \frac{3}{4\pi} \cdot \frac{\langle \sigma v \rangle}{m_\chi^2} \cdot \frac{M^2}{R_{cl}^3} \cdot S. \quad (2.58)$$

Here as well as in Eq. (2.56) the variable r is the distance from the centre of the clump. The WIMP mass is characterised by m_χ and the clump mass by M. The value of the parameter S is constant and depends on the density profile of the DMC. In the trivial case of a homogenous sphere the value of S is equal to unity. Then the annihilation signal from a DMC which can be observed at the position of the Earth is given by

$$\Phi_{DM,clump} = \frac{1}{4\pi} \int_0^\pi d\theta \sin(\theta) \int_0^{r_{max}(\theta)} dr \frac{2\pi r^2}{r^2} \int_{M_{min}}^{M_{max}} dM \int_{R_{min}}^{R_{max}} dR \, n_{cl}(R, M, R_{cl}) \dot{N}_{cl}(M, r), \quad (2.59)$$

where n_{cl} is the number density of DMCs in the Galactic halo. The variable R is the galactocentric distance which varies between R_{min} and R_{max}. With the knowledge of the virial radius R_{vir} of the MW, which is defined as the distance from the GC where the accumulated density of the MW is 200 times the critical density of the Universe, the maximal distance R_{max} can be calculated with law of cosine and the galactocentric distance of the Sun.
A simple approximation can be done by the introduction of a so-called boost factor B_l which characterises the increased luminosity of a clump with respect to the annihilation signal of the smoothly distributed DM component. Assuming that the average clump luminosity is equal in all directions the boost factor is independent of the direction of observation and the number density of the DMCs can be parametrised by a radially dependent distribution. In this case the signal of the clumpy DM component can be estimated analogously to the signal of the diffuse component

$$\Phi_{DM,clump}(E, \psi, \Delta\Omega) = \frac{\langle \sigma v \rangle}{4\pi} \cdot \sum_f \frac{dN_f}{dE} b_f \cdot \frac{1}{\Delta\Omega} \cdot B_l \cdot \int_{\Delta\Omega} \int_{los} n_{cl}(R) \, dl_\psi, \quad (2.60)$$

where B_l is the boost factor which then depends on the clump density profile and the primordial spectrum of density fluctuations.
In the early stage of structure formation the DMCs were destroyed by the hierarchical clustering process but nowadays this process is negligible. Today clumps are mainly destroyed by gravitational forces produced by massive objects in the Galactic disc or the potential of the entire MW. Since a DMC is a wide spread object the gravitational attraction towards a

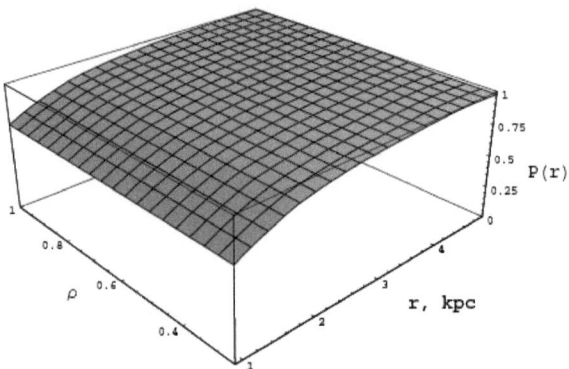

Figure 2.15: Survival probability for DM clumps with respect to the Galactocentric distance r and the averaged clump density. Figure taken from [10].

massive object is different for the nearby and far away end of the clump. This effect leads to different acceleration velocities of the DM particles inside the clump and destroys the clump. That is why this effect is called tidal disruption. During the disruption process the outer parts of the clumps are stripped away and form so-called tidal streams. The tidal disruption of clumps is reflected in a constant boost factor (like in the case of no tidal desruption) multiplied with a radially dependent survival probability which was calculated in [10]. This survival probability of DMCs is crucially dependent on the value of x_c. For $x_c \gg 0.05$ the survival probability at the GC P(0) is about zero which means that the clump is completely destroyed. For smaller values of $x_c \ll 0.05$ only the outer parts of the clump are stripped away and the core of the clump remains. In this case the central value of $P(0) = 1$ since the major contribution of annihilation flux of a clump stems from the core. In Figure 2.15 the survival probability is shown for $x_c = 0.05$. Then the DMA flux of the clumpy DM component is given by

$$\Phi_{DM,clump}(E, \psi, \Delta\Omega) = \frac{\langle \sigma v \rangle}{4\pi} \cdot \sum_f \frac{dN_f}{dE} b_f \cdot \frac{1}{\Delta\Omega} \cdot B_l \cdot \int_{\Delta\Omega} \int_{los} P(R) \, n_{cl}(R) \, dl_\psi. \quad (2.61)$$

As the analysis of the annihilation signal of different DM density distributions for the smooth and the clumpy DM component is crucially dependent on the Galactic background radiation produced by cosmic radiation. The relevant background processes are considered in the following section.

2.7 Background Processes

In the previous section the gamma radiation flux produced by the different DM components was explained. However, gamma radiation is not only produced by DMA but also by the interaction of cosmic radiation (CR) with the interstellar medium (ISM). Cosmic rays are the highest energetic particles yet known. Their sources and acceleration processes are still subject of current discussions. The primary particles of the CR are mainly protons with small contributions of electrons and helium nuclei. These particles are influenced by the magnetic and the radiation field of the Galaxy, so they do not propagate along a straight line but their trajectory looks like a staggered course. The energy spectrum of the diffuse gamma radiation at the Earth is influenced by the gamma rays produced by the interaction of CR protons and electrons with the ISM. In the standard model of particle physics the main interaction processes are

$$\begin{aligned} e + \gamma &\rightarrow e' + \gamma', \\ e + p &\rightarrow e' + p' + \gamma, \\ p + p &\rightarrow \pi^0 + X \rightarrow \gamma\gamma + X. \end{aligned} \quad (2.62)$$

The main contribution of the Galactic gamma rays is produced in the pion decay (pp) process. There high energetic protons or nuclei from cosmic radiation interact with protons of the interstellar gas. The shape of the energy spectrum of this contribution is consequently dependent on the cosmic ray energy spectrum and the density distribution of the ISM. These distributions are locally measured [75]. If the cosmic ray spectrum at the Earth is taken to be representative for the entire Galaxy, called conventional model, the energy dependence of the photons of the pp process is equal in all directions. The contributions of the inverse Compton scattering process (eγ) and the Bremsstrahlung (ep) are small compared to the pion decay process. The spectral shapes of the different components are known from accelerator experiments and implemented in Monte Carlo generators like `Pythia` [77]. However, the gamma ray spectrum produced with this model is not hard enough to explain the measured energy spectrum of the diffuse Galactic gamma radiation. In order to explain the measured data without introducing an additional component to the gamma ray spectrum the local cosmic ray spectrum have to be harder which means that the locally measured cosmic ray spectrum is not representative for the Galaxy. Such a model is called optimised model [78]. For a harder gamma ray spectrum the cosmic ray spectrum was modified in the low energy region because in the high energy region the slope of the cosmic ray spectrum has to be compatible with measured data. In the low energy region the flux of charged particles is influenced by the magnetic field of the Sun. Due to this effect the slopes in the low energy region can be different from the slopes predicted by the conventional model and a harder spectrum can be obtained.

In [16] both background models were used to analyse the gamma ray energy spectrum. There the spectral shape of the gamma ray spectrum was fixed while the normalisation of the gamma ray components was left free. It was found that even in case of an optimised model the fit to the data was improved by introducing a gamma ray signal from DMA. The conventional model was found to be consistent with a WIMP mass in the range from 50 to 60 GeV while the upper limit is increased to 100 GeV for a optimised model.

In the present analysis the propagation code `Galprop` [79] is used to calculate the energy spectrum of the background radiation in different sky regions. There the propagation of CR particles through the Galaxy is calculated by the numerical solution of the transport equation which is given by

$$\frac{\partial \psi}{\partial t} = q(\vec{r},p) - \frac{1}{\tau_f}\psi - \frac{1}{\tau_r}\psi + \vec{\nabla}\cdot\left(D_{xx}\vec{\nabla}\psi - \vec{V}\psi\right) + \frac{\partial}{\partial p}p^2 D_{pp}\frac{\partial}{\partial p}\frac{1}{p^2}\psi - \frac{\partial}{\partial p}\left[\dot{p}\psi - \frac{p}{3}\left(\vec{\nabla}-\vec{V}\right)\psi\right]. \quad (2.63)$$

There $\psi = \psi(\vec{r},p,t)$ is the time-dependent particle density in phase space, $q(\vec{r},p)$ describes the source term for CRs, D_{xx} and D_{pp} describe the diffusion coefficient in space and momentum space. The convection velocity is given be $\vec{V}$, the energy loss rate of the particles is represented by $\dot{p}$ and the time scales for fragmentation and radioactive decay are given by τ_f and τ_r. The rigidity dependence of the spatial diffusion coefficient is given by

$$D_{xx} = \beta \cdot D_0 \cdot \left(\frac{p}{p_0}\right)^\delta, \quad (2.64)$$

where β is the particle velocity and ρ_0 describes a possible break in δ. In case of reacceleration is taken into account the relation between the diffusion coefficient in space and in momentum space is given by

$$D_{pp}\cdot D_{xx} = \frac{4\, p^2\, v_A^2}{3\delta(4-\delta^2)(4-\delta)w}. \quad (2.65)$$

The parameter v_A is the Alfven velocity which describes the velocity of waves in the cosmic ray plasma. The level of turbulence in the plasma is given by w and is equal to the ratio of the magnetohydrodynamic wave energy density to the magnetic field density. If convection is taken into account the particle velocity V is assumed to increase linearly with vertical height z which corresponds to cosmic ray driven magnetohydrodynamic wind models [80]. The source term for protons and electrons is estimated by a power law

$$q(\vec{r},p) = q_0(\vec{r}) \cdot \left(\frac{p}{p_0}\right)^{\gamma_{p,e}} \quad (2.66)$$

with a possible break at p_0 in the spectal index $\gamma_{p,e}$. For the nucleon energy losses ionisation and Coulomb interactions are included while for electrons bremsstrahlung, inverse Compton scattering and synchrotron radiation is taken into account. The transport equation in Eq. (2.63) is numerically solved in cylindrical coordinates. The convection is considered up to a maximal vertical boundary z_h. Beyond this boundary free escape of the CR is assumed. The propagation equation is successively solved beginning with the heaviest nuclei to the lightest nuclei including all secondary source functions. Subsequently, the propagation of electrons and positrons is calculated. The energy spectra of the gamma radiation produced by bremsstrahlung and inverse Compton scattering is then calculated with the propagated particles and the gas distribution and radiation fields used for their propagation. The gamma radiation produced by the pp process is calculated with the propagated energy spectra of the protons and helium with the interstellar gas. The values of the conventional model used for this analysis are summarised in the datacard file presented in Appendix A.

A further contribution to the background gamma radiation stems from extragalactic sources. This radiation is likely produced by active galactic nuclei (AGN), quasars or blazars of nearby galaxies. It increases relatively with increasing Galactic latitudes since the influence of the Galactic disc is decreasing for higher latitudes. However, the extragalactic background radiation (EGBR) is different for positive and negative latitudes. The shape of its contribution to the energy spectrum of the gamma radiation is unknown since every extragalactic object has individual properties. In the present analysis the EGBR is calculated according to an iterative method including DMA [81]. There the gamma ray skymap is divided in several regions in order to get various values of the gamma ray fluxes. The Galactic plane and the Galactic pole are excluded since in the disc the EGBR is negligibly small and insufficient amount of data at the Galactic poles. For each energy bin the measured gamma radiation flux is plotted against the expected gamma ray flux. The approximation of this plot shows a linear dependence with a slope equal to unity if a realistic background radiation model is chosen. Then the extrapolation to zero provides the EGBR per energy bin. The expected gamma radiation flux depends on the background radiation model and the spectrum of the gamma radiation from DMA. Therefore, the EGBR is different for different background radiation models and different WIMP masses. For a rough division of the sky a slight dependence of the DM density distribution is expected since variations are averaged in this case.

3
Constraints on the Dark Matter density distribution from astronomical observations

The total Galactic density distribution is subject to different astronomical observations. The rotation velocity distribution in the Galaxy depends on the radial density distribution while determination of the surface density at the Sun depends on the density decrease perpendicular to the Galactic disc. Therefore, a combination of these measurements provides an indication for the DM density distribution in the Galactic halo.

In the last years new results from astronomical data became available which can be used to constrain the density distribution of the MW. For this reason a reconsideration of the distribution of the luminous matter and DM in the Galaxy becomes worthwhile. First the astronomical observations and the the parametrisation of the different matter contribution (luminous matter and DM) are explained. After that the astronomical constraints are presented. At the end of the chapter an additional DM substructure composed of two circular rings is examined and a discussion of the results is given.

3.1 Astronomical observations

The Galactic DM distribution is constrained by astronomical observations. In the following sections the astronomical constraints used in this analysis are explained.

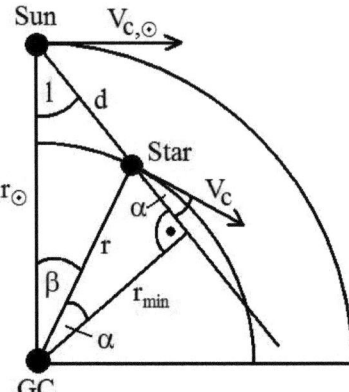

Figure 3.1: Schematic picture of the Galaxy. The values r and V_c represent the Galactocentric distance and the rotation velocity of the observed star while r_{min} describes the minimal distance to the line-of-sight at which the light from the star travels to the Sun. The Galactocentric distance of the Sun is $r_\odot$ and its rotation velocity is $V_{c,\odot}$. The distance from the star to the Sun is d and the Galactic longitude at which the star is visible on the sky is given by l. The value of β is the azimuthal angle.

3.1.1 Rotation velocity and Galactocentric distance to the Sun

The rotation velocity of the Solar System around the GC and its Galactocentric distance is determined by the observation of the kinematics of stars. Assuming that Galactic objects rotate on circular orbits around the GC and that the observed stars lie within the orbit of the Sun the observation is schematically shown in Figure 3.1. The Galactocentric distance to the Sun is given by $r_\odot$ and its rotation velocity is given by $V_{c,\odot}$. The distance from the Sun to the observed star is represented by d while the Galactocentric distance of the star and its rotation velocity is given by r and V_c. From Figure 3.1 the following relations can be derived

$$r \cdot \sin(\alpha) = r_\odot \cdot \cos(l) - d, \qquad (3.1)$$
$$r \cdot \cos(\alpha) = r_\odot \cdot \sin(l), \qquad (3.2)$$
$$r_\odot = d \cdot \cos(l) + r \cdot \cos(\beta). \qquad (3.3)$$

3.1 Astronomical observations

The radial and tangential velocity of the star with respect to the Sun is given by

$$V_r = V_c \cdot \cos(\alpha) - V_{c,\odot} \cdot \sin(l), \tag{3.4}$$
$$V_t = V_c \cdot \sin(\alpha) - V_{c,\odot} \cdot \cos(l). \tag{3.5}$$

The angular velocity of the solar system $\Omega_\odot = V_{c,\odot}/r_\odot$ and of the observed star $\Omega = V_c/r$ in combination with the Eqs. (3.2) and (3.3) result in the radial and tangential velocity of the star

$$V_r = (\Omega - \Omega_\odot) \cdot r_\odot \cdot \sin(l), \tag{3.6}$$
$$V_t = (\Omega - \Omega_\odot) \cdot r_\odot \cdot \cos(l) - \Omega \cdot d. \tag{3.7}$$

Assuming the observed star is in the vicinity of the Sun, $d \ll r_\odot$ or $|r - r_\odot|$ is small, the angular velocity of the star can be approximated using the Taylor expansion $\Omega \approx \Omega_\odot + (\partial \Omega / \partial r) \cdot (r - r_\odot)$. Then Eq. (3.3) subsequently changes to $r_0 = d \cdot \cos(l) + r$. Using this equation, the Taylor expansion of Ω and the additional theorems of the trigonomical functions the expression for the velocities of the star result in

$$V_r = A \cdot d \cdot \sin(2l), \tag{3.8}$$
$$V_t = A \cdot d \cdot \cos(2l) + B \cdot d. \tag{3.9}$$

A and B are the so-called Oort constants which are given by

$$A = -\frac{1}{2} \cdot \left[\left(\frac{\partial V_c}{\partial r} \right)_{r_\odot} - \frac{V_{c,\odot}}{r_\odot} \right], \tag{3.10}$$

$$B = -\frac{1}{2} \cdot \left[\left(\frac{\partial V_c}{\partial r} \right)_{r_\odot} + \frac{V_{c,\odot}}{r_\odot} \right]. \tag{3.11}$$

These values represent the azimuthal shear (A) and the vorticity (B) of the velocity field generated by closed orbits. The most precise determination and therewith the normalisation of the rotation curve is obtained from the Oort constants, which can be determined from the precise distances and velocities of nearby stars [48, 82, 83]. The experimental values of the Oort constants

$$A \approx 14.4 \pm 1.2 \text{ km s}^{-1} \text{ kpc}^{-1} \tag{3.12}$$
$$B \approx -12.0 \pm 2.8 \text{ km s}^{-1} \text{ kpc}^{-1}, \tag{3.13}$$

have been taken from [84]. The difference $A - B = v_\odot/r_\odot$ defines the local normalisation of the rotation curve. The combination A-B can be more precisely determined than the individual constants. [84] found $A - B = v_\odot/r_\odot = 27 \pm 2.5$ km s^{-1} kpc^{-1}. Using the proper motion of the black hole in the Galactic centre (Sgr A*) [85] found

$$A - B = v_\odot/r_\odot = 29.45 \pm 0.15 \text{ km s}^{-1} \text{kpc}^{-1}, \tag{3.14}$$

which is in excellent agreement with recent measurements of parallaxes using the Very Large Baseline Interferometry (VLBI) [86], which yield $A - B = v_\odot/r_\odot = 30.3 \pm 0.9$ km s^{-1} kpc^{-1}. A further interesting property of the RC is its slope at the position of the Sun. This value is strongly connected to the values of A and B and was determined in [87]. The slope of the RC at the Sun was found there to be

$$Slope_{\odot,RC} = \left.\frac{\partial ln(v)}{\partial ln(r)}\right|_{r_\odot} = -\frac{A+B}{A-B} = -0.006 \pm 0.016. \quad (3.15)$$

For the sake of completeness it is mentioned that a generalisation to an non-axisymmetric density distribution leads to two additional constants (C and K) which describe the radial shear (C) and the local divergence (K) of the velocity field [88]. In this analysis a circular (or a slightly elliptical) rotation within the Galactic disc is assumed. In this case C ≈ K ≈ 0. From the observation of 28 stars orbiting Sgr A*, which is considered to be the centre of the Galaxy because of its small own velocity, over nearly 16 years the distance between the Sun and the GC has been determined to [89]:

$$r_\odot = 8.33 \pm 0.35 \text{ kpc}, \quad (3.16)$$

in agreement with [90]. With this Galactocentric distance a rotation velocity of the Sun of

$$v_\odot = 244 \pm 10.2 \text{ km s}^{-1}, \quad (3.17)$$

can be derived using Eq. (3.14). This speed determines the mass of the Galaxy inside the solar radius.

The Galactocentric distance $r_\odot$ and the rotation velocity $v_\odot$ of the Sun are important values for the determination of the kinematics of Galactic objects which is considered in the next section.

3.1.2 Rotation curve of the Milky Way

Each object, which is bound to the Milky Way, is orbiting around the Galactic centre where the gravitational potential is strongest. Most of the stars and the interstellar medium are located in the Galactic disc which rotates with a peculiar velocity distribution $v(r)$ as well. This velocity distribution is called the rotation curve (RC) of the Milky Way.

The comparison of the estimated velocity distribution with the rotation curve measured for the Galactic disc of the Milky Way is an important check for the DM density profile. However, the measurement of the RC of the Milky Way is not trivial since the Sun is also located in the Galactic disc and all observations are done from this position. Therefore, the Galactocentric distance $r_\odot$ and the rotation velocity $v_\odot$ of the Sun has to be known in order to determine the distance of a Galactic object to the GC and to define the standard frame of rest. The determinations of $r_\odot$ and $v_\odot$ were already discussed in the last section.

The rotation velocity is measured for clouds of interstellar gas, like hydrogen and carbon monoxide, and stars. In case of circular rotation the velocity of gas clouds can be determined by the observation of their emitted radiation. For gas clouds within the solar circle the so-called tangent point method is used. Therein the line-of-sight on which the light emitted by the gas cloud travels to the Earth is tangential to the rotation circle of the gas. The rotation velocity can then be determined with the measurement of the Doppler shift of the spectrum of the emitted light. This method provides accurate values for the velocity distribution in the inner Galaxy.

The determination of the outer rotation curve is rather difficult because a tangent point to the rotation orbit of the gas cloud cannot be defined there. In this case the rotation velocity is usually determined by the measurement of the distance from the observed gas cloud to the Earth and its angular rotation velocity. However, the measurement of distances to Galactic objects is not easy since stars, nebulae, etc. are only visible on the sky in two dimensions. The third dimension, the distance to the object, remains hidden. In order to obtain the distance different procedures are used. For small distances of a few hundred parsecs the trigonometric, photometric or spectrometric parallax is used. In this case the small periodic shift in the apparent location of the observed object which results from the changing location of the Earth as it orbits around the Sun is determined. Another method to determine the distance is the so-called distance modulus which is the difference of the apparent luminosity and the absolute luminosity of an object. Therefore, a catalogue of objects with known properties, called cosmic distance ladder, is needed. These objects are called standard candles. For example, Cepheid variables or RR Lyrae stars with known absolute magnitudes are used as standard candles. This methods for the determination of the distance to the Earth unfortunately yield large uncertainties for larger distances which is reflected in the determination of the rotation velocity for the outer Galaxy.

In 1992 a new method for the determination of the outer rotation curve was introduced [91]. Assuming that the rotation of the Galactic disc is circular and that the vertical distribution of the gas only depends on the Galactocentric distance the distance to a rotating gas cloud is obtained by the measurement of the vertical half width of the H_2 distribution with this method. However, in this case a detailed map of the vertical height of the gas distribution is necessary since measurements have shown that the scale height is different for the northern and the southern hemisphere [20]. This method works up to a radius of $r \approx 2 \cdot r_\odot$. For larger radii the vertical H_2 distribution is too frayed.

The velocity of a Galactic object is estimated from the matter density model by the calculation of the gravitational potential. Assuming that the object is circularly rotating around the GC the rotation velocity is given by the equality of the centripetal and the gravitational force

$$\frac{mv^2}{r} = m \cdot \frac{d\Phi}{dr}, \qquad (3.18)$$

where v is the rotation velocity at the galactocentric distance r. The gravitational potential Φ is given by the Poisson equation

$$\Delta\Phi = 4\pi G\rho(r), \tag{3.19}$$

where G is the gravitational constant and $\rho(r)$ the density distribution of the Galaxy. Integrating the density distribution twice yields the corresponding gravitational potential. In spherical coordinates one finds:

$$\Phi(r,\theta,\phi) = -\int_0^\infty r'^2 dr' \int_{-1}^1 dcos\theta' \int_0^{2\pi} d\phi' \frac{\rho(r',\theta',\phi')}{\sqrt{r^2 + r'^2 - 2rr'\sin\theta\sin\theta'\cos(\Delta\phi) - 2rr'\cos\theta\cos\theta'}}, \tag{3.20}$$

where $\Delta\phi = \phi - \phi'$. For the MW two different rotation velocity distributions have been measured: the radial dependence of rotation speed within the Galactic disc and the rotation velocity of halo objects far away from the Galactic plane.

Rotation curve in the Galactic disc

For the calculation of the gravitational potential in the Galactic plane the polar angle θ in Eq. (3.20) has to be set to $\pi/2$. In this case the last term of the square root in the denominator vanishes and the potential is given by

$$\Phi(r,\pi/2,\phi) = -\int_0^\infty r'^2 dr' \int_{-1}^1 dcos\theta' \int_0^{2\pi} d\phi' \frac{\rho(r',\theta',\phi')}{\sqrt{r^2 + r'^2 - 2rr'\sin\theta'\cos(\Delta\phi)}}. \tag{3.21}$$

Following Eq. (3.18) the rotation velocity within the Galactic plane is given by the derivation of the gravitational potential with respect to the Galactocentric distance

$$\frac{v^2}{r} = \frac{d\Phi(r)}{dr} = \int_0^\infty r'^2 dr' \int_{-1}^1 dcos\theta' \int_0^{2\pi} d\phi' \frac{\rho(r',\theta',\phi')(r - r'\sin\theta'\cos(\Delta\phi))}{(r^2 + r'^2 - 2rr'\sin\theta'\cos(\Delta\phi))^{3/2}}. \tag{3.22}$$

For the RC within the Galactic disc a combination of different measurements with different tracers was summarised in [92]. The experimental data were scaled to $v_\odot = 244$ km s^{-1} at a Galactocentric distance of 8.3 kpc, as shown in Figure 3.2a. The rotation velocities in the very inner part of the Galaxy (radii smaller than 2.5 kpc) are obtained from observations of HI regions and their associated molecular clouds (CO) using the tangent point method [93]. Further data on the velocity distribution within the Sun's orbit is adapted from [94, 95] also by tangent point observations of HI and CO regions. Beyond the solar circle the velocity distribution is obtained by the observation of HII regions, planetary nebulae and stars [95, 96]. These data sets have high experimental uncertainties by virtue of uncertain determinations of the distance to the observed objects. Later publications provide velocity distributions up to radii of about 20 kpc [97, 98]. There the RC beyond the orbit of the solar system

3.1 Astronomical observations

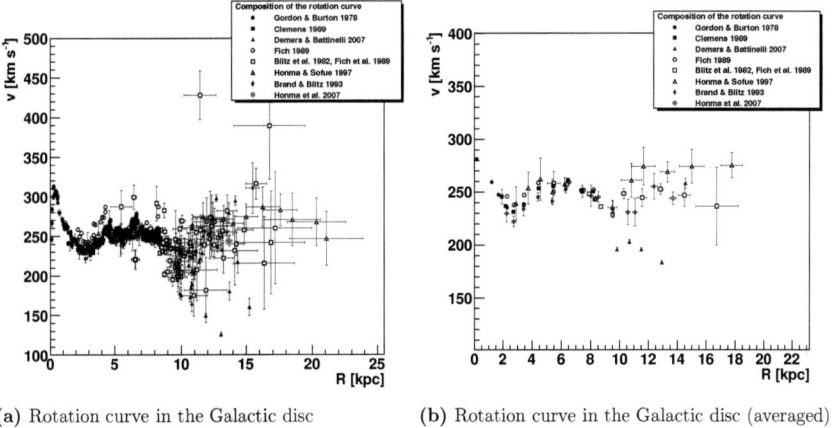

(a) Rotation curve in the Galactic disc

(b) Rotation curve in the Galactic disc (averaged)

Figure 3.2: Velocity distribution in the Galactic disc of the MW.

is obtained by using Merrifield's method [91] which leads to smaller experimental errors. Unfortunately, detailed observations of the RC in [98] are not included in the combination of [92]. In the present analysis these data are included. On the other hand the observation of carbon stars in the radial region of 9 to 15 kpc in [99] are excluded since the observed stars show very low rotation velocities which probably results from their common origin in the Canis Major star overdensity.

The rotation velocity was averaged in 18 radial bins from the GC up to a radius of 22 kpc, as shown in Figure 3.2b. The different data sets as well as their averaged velocities in the different radial bins are presented in Appendix B. The measured velocity distribution shows a strong increase in the inner part of the Galaxy which presumably results from the dense core of the Galaxy. In the region between 6 and 10 kpc a decrease of the rotation velocity is observed. Then the rotation speed increases again. Such a peculiar change of slope cannot be explained by a smoothly decreasing DM density profile, but needs an additional ringlike DM substructure, e.g. the infall of a dwarf Galaxy, as mentioned in [16]. The vertical thickness of this structure is of the order of 1 kpc, so it should not show up for halo stars well above this height. The rotation of stars in the halo around the GC is explained in the following section.

Rotation curve in the halo

The rotation of halo stars around the GC does not proceed within the Galactic disc. The plane in which a star is orbiting the GC is inclined with respect to the Galactic plane.

The kinematics of a large sample of roughly 2400 blue horizontal-branch (BHB) tracer halo stars as detected in Sloan Digital Sky Survey (SDSS), with Galactocentric distances up to about 60 kpc and vertical heights of $z > 4$ kpc, was recently analysed [100]. There the halo star distribution from N-body simulations of the Galaxy with an NFW profile was used to connect the observable values — line-of-sight velocity and distance — to the circular velocity $v(r) = \sqrt{r\, \partial\Phi/\partial r}$. In order to estimate the rotation curve of halo stars the gravitational potential as given in Eq. (3.20) has to be deviated according to the Galactocentric distance of the halo star, which results in

$$\frac{v^2}{r} = \frac{d\Phi(r)}{dr} = \int_0^\infty r'^2 dr' \int_{-1}^1 d\cos\theta' \int_0^{2\pi} d\phi' \frac{\rho(r', \theta', \phi')(r - r'\sin\theta'\sin\theta\cos(\Delta\phi) - r'\cos\theta\cos\theta')}{(r^2 + r'^2 - 2rr'\sin\theta'\sin\theta\cos(\Delta\phi) - 2rr'\cos\theta\cos\theta')^{3/2}}.$$
(3.23)

The sample of stars in [100] is observed for Galactic latitudes higher than about 45°. For this reason the halo star velocity curves are calculated for the latitudes 45° and 80°. Then these rotation curves are averaged for comparision with experimental data.

3.1.3 Mass of the Galaxy

In general, the total Galactic mass is measured indirectly either via the kinematics of distant halo tracer stars, satellite galaxies or the vertical scale height of the gas distribution of the Galactic disc.

The total mass of a galaxy is conventionally defined as the mass within the so-called virial radius. At this radius the total mass of the accumulated density of the MW is equal to the mass of a homogeneous sphere with a constant density of 200 times the critical density of the Universe.

In [101] the mass of the MW was estimated from measurements of the radial velocities of 27 globular clusters and satellite galaxies for Galactocentric distances $r > 20$ kpc, using a Bayesian likelihood method and a spherical halo mass model with a truncated radius. They found a mass of the Galaxy within 50 kpc of $M(50\text{ kpc}) = 5.4^{+0.2}_{-3.6} \cdot 10^{11}\ M_\odot$ and a total mass of $M_\text{tot} = 1.9^{+3.6}_{-1.7} \cdot 10^{12}\ M_\odot$. A similar analysis was performed with more tracer stars by [102]; they found $M_\text{tot} = 2.5^{+0.5}_{-1.0} \cdot 10^{12}\ M_\odot$. These measurements used a simple parametrisation of the potential. Analyses using an NFW profile for the DM distribution usually obtain a lower total mass because of the steeper fall-off of the density profile at large distances.

Using a large sample of 2400 BHB stars from the SDSS in the halo ($z > 4$ kpc, $r < 60$ kpc) and comparing the results with N-body simulations using an NFW profile find

$$M_{r<60\text{ kpc}} = 4.0 \pm 0.7 \cdot 10^{11} M_\odot,$$
(3.24)

which corresponds to $M_\text{tot} = 1.0^{+0.3}_{-0.2} \cdot 10^{12}\ M_\odot$ [100]. To get this total mass adiabatic contraction [103] was allowed in which case the concentration parameter c, which is the ratio of the virial radius and the scale radius of the DM halo profile, was taken as a free parameter in

Contribution	Surface density [$M_\odot$ pc^{-2}]	Reference
Visible stars	35 ± 5	[109]
	27	[110]
	30	[111]
Stellar remnants	3 ± 1	[112]
Interstellar gas	8 ± 5	[113]
	13 - 14	[114]
Total	35-58	

Table 3.1: Contributions to the local surface density of baryonic matter. The total values in the last row include $\pm 1\sigma$ errors.

the fit. This parameter turned out to be low ($c = 6.6^{+1.8}_{-1.6}$; if adiabatic contraction is ignored, the concentration parameter is around 12 and the mass decreases to $M_{tot} = 0.9 \cdot 10^{12} \, M_\odot$, which is well within the errors. Similar mass values were found from the measurement of the velocity dispersion in [104], although with larger errors. These measurements are consistent with the favoured ΛCDM model in [105] where the virial mass is $1 \cdot 10^{12} \, M_\odot$. The value from Eq. (3.24) will be used in the analysis.

3.1.4 Local surface density and Oort limit

In addition to the velocity distributions in the Galactic disc and the DM halo the total surface density $\Sigma_{\odot,\text{tot}} = \Sigma_{\odot,\text{vis}} + \Sigma_{\odot,\text{DM}}$ and the total density $\rho_{\odot,\text{tot}} = \rho_{\odot,\text{vis}} + \rho_{\odot,\text{DM}}$ at the Solar position is used to constrain the density model. These are obtained as follows.
Integrating the density along the vertical direction within $\pm z$ from the Galactic plane yields the surface density

$$\Sigma_{def}(<|z|) = \int_{-z}^{z} \rho(z') \, dz'. \qquad (3.25)$$

First the surface density of the visible matter is considered. Its experimental value is obtained by the observational determination of the luminous matter density in the vicinity of the Sun. The total value results from the summation of the different contributions to the luminous matter — the stellar population, stellar remnants and the interstellar gas. A summary of the different measurements was given in [106]. The surface density of the baryonic matter lies between 35 and 58 $M_\odot$ pc^{-2} (Table 3.1), which agrees with the estimation 48 ± 9 $M_\odot$ pc^{-2} [107, 108]. On the contrary to star counts dynamical determinations show even larger values. In this case the surface density is obtained by the comparison of the star velocities with the predictions of either a mass model of the Galaxy or a parametrisation of the gravitational potential at the position of the Sun. In [107] the local surface density was determined from a parametrisation of a mass distribution to be

$$\Sigma(< 1.1 \text{ kpc}) = 71 \pm 6 \text{ M}_\odot \text{ pc}^{-2} \qquad (3.26)$$

while in [108] the modeling of the vertical gravitational potential resulted in $\Sigma(< 1.1 \text{ kpc}) = 74 \pm 6 \text{ M}_\odot \text{ pc}^{-2}$. The most recent determination of the surface density by [115] is consistent with both measurements but allows somewhat larger errors $\Sigma(< 1.1 \text{ kpc}) = 68 \pm 11 \text{ M}_\odot \text{ pc}^{-2}$.

In case of modeling the density distribution of the Galaxy the surface density is easily determined by the integration in Eq. (3.25). If the graviational potential is parametrised the surface density can be determined using Poisson's equation (see Eq. (3.19)) which relates the density to the second derivation of the gravitational potential

$$\Delta\Phi = \frac{1}{r}\frac{\partial}{\partial r}\left(r\frac{\partial\Phi}{\partial r}\right) + \frac{1}{r^2}\frac{\partial^2\Phi}{\partial\varphi^2} + \frac{\partial^2\Phi}{\partial z^2}. \qquad (3.27)$$

Consequently, the surface density at the position of the Sun is obtained by the first derivative perpendicular to the Galactic disc of the gravitation potential at the Sun. This corresponds to the gravitational force orthogonal to the disc.

$$\begin{aligned}
\Sigma(<|z|) &= \frac{1}{2\pi G}\int_0^z \Delta\Phi\, dz' \\
&= \frac{1}{2\pi G}\int_0^z \left(\frac{1}{r}\frac{\partial}{\partial r}\left(r\frac{\partial\Phi}{\partial r}\right) + \frac{1}{r^2}\frac{\partial^2\Phi}{\partial\varphi^2} + \frac{\partial^2\Phi}{\partial z'^2}\right) dz' \\
&= \frac{1}{2\pi G}\left(\frac{\partial\Phi}{\partial z} + \int_0^z \frac{1}{r}\frac{\partial}{\partial r}\left(r\frac{\partial\Phi}{\partial r}\right) + \frac{1}{r^2}\frac{\partial^2\Phi}{\partial\varphi^2}\right)dz' \\
&= \frac{1}{2\pi G}\left(\frac{\partial\Phi}{\partial z} + 2\cdot(A^2 - B^2)\cdot |z|\right) \\
&\approx \frac{1}{2\pi G}\left(\frac{\partial\Phi}{\partial z}\right).
\end{aligned} \qquad (3.28)$$

Assuming that the rotation orbits are approximately circular and the absolute values of the Oort Constants A and B are approximately equal the surface density is given by the derivation of the gravitational potential in vertical direction.

In [116] the vertical potential was determined by a least-square fit of

$$\phi(z) = 2\pi G \cdot \left(\Sigma \cdot \left(\sqrt{z^2 - D^2} - D\right) + \rho_{eff}\cdot z^2\right) \qquad (3.29)$$

in which D represents the scale height of luminous matter in the Galactic disc, Σ is the surface density and ρ_{eff} is the density contribution of a spherical DM halo. In [115] a good

fit is obtained for values of ρ_{eff} within the range of 0 to 0.02 $M_\odot$ pc^{-3}.
The first integration in z of Eq. (3.29) gives the total surface mass density within $\pm z$ from the Galactic disc. The second integration in z gives the total mass density depending on z. Therefore the total mass density of the Galactic disc at the position of the Sun is given by

$$\rho_{tot}(z = 0 \text{ kpc}) = \frac{\Sigma}{2D} + \rho_{eff}. \tag{3.30}$$

Jan Oort proposed and performed an interesting measurement of the local matter density: from the number of stars as a function of their height above the disc one can obtain the local gravitational potential, which is directly proportional to the mass in the plane of the MW. Using the precise measurements from the Hipparcos satellite one obtains for the local mass density

$$\rho_{\odot,tot}(z = 0 \text{ kpc}) = 0.102 \pm 0.010 \ M_\odot \ \text{pc}^{-3}, \tag{3.31}$$

which includes visible and dark matter [108]. This value is called Oort limit, since it represents the lowest value for the local density. It was determined from the precise star counts and velocity measurements in a volume of 125 pc around the Sun. In [117] the vertical potential at slightly larger distances (a vertical cylinder of 200 pc radius and an extension of 400 pc out of the Galactic plane) was analysed. For the dynamical estimation of the local volume density they obtain the same value with a smaller error: $\rho_{\odot,\text{tot}}(z=0) = 0.100 \pm 0.005 \ M_\odot \ \text{pc}^{-3}$. To be conservative, we will use the value from Eq. (3.31).

Gas flaring

The observation of the gas distribution is performed by the measurement of the radiative emission of atomic hydrogen. The 21 cm line is a perfect tracer for this purpose since under most circumstances the interstellar medium is transparent at this peculiar wavelength. The results of the measurement of the spatial distribution of the 21 cm emission can be found in [20]. Therein the half-width-half-maximum (HWHM) of the vertical decrease of the interstellar gas distribution was measured for all Galactic longitudes, and a difference for the northern and southern hemisphere of the Galactic disc was discovered. However, the most interesting feature in the radial dependence of the HWHM was a flattening in the radial region between 14 kpc and 20 kpc instead of an increase. This effect can be explained with a DM ring at the Galactocentric distance of about 13 kpc [20]. There the gas flaring results were described with an exponentially decreasing density distribution for the Galactic disc with an additional DM ring of a mass of about $2.0 \cdot 10^{10} \ M_\odot$.

In analogy to the barometric equation the vertical decrease of the interstellar hydrogen can be parameterised by [116]

$$\rho(z) = \rho_{\odot,tot} \cdot \exp\left(-\frac{\Phi(z)}{\omega^2}\right). \tag{3.32}$$

The vertical gravitational potential is given by $\Phi(z)$ and the vertical velocity dispersion of the gas is ω. The experimental determination of the velocity dispersion yielded values of

7 ± 1 km s^{-1} [118] and 8 ± 1 km s^{-1} [119].

3.2 Parametrisation of the density distribution of the Milky Way

The assumption of the different parts of the density distribution in the MW is the most important part of the analysis. In this section the parametrisation of the luminous matter and the DM density distribution of the Galaxy is explained.

3.2.1 Luminous matter density

The density distribution of the luminous matter of a spiral galaxy can be splitted into two parts, the Galactic disc and the Galactic bulge. The parametrisation of the density distribution of the bulge is adapted from [120]

$$\rho_b(r,z) = \rho_b \cdot \left(\frac{m}{r_{0,b}}\right)^{-\gamma_b} \cdot \left(1 + \frac{\tilde{r}}{r_{0,b}}\right)^{\gamma_b - \beta_b} \exp\left(-\frac{\tilde{r}^2}{r_t^2}\right), \quad \text{where } \tilde{r}^2 = \sqrt{x^2 + y^2 + (z/q_b)^2}.$$

For a good description of the RC near the GC the parameters of the bulge profile are found to be $\gamma_b = \beta_b = 1.6$, $q_b = 0.61$, $r_t = 0.6$ kpc, $r_{0,b} = 0.7$ kpc and $\rho_b = 360$ GeV cm^{-3}, which corresponds to approximately 9.5 M$_\odot$ pc^{-3}. This high density may be influenced by the presence of black holes. At least one black hole with a mass of about $4.0 \cdot 10^6$ M$_\odot$ has been observed in the GC [121].

The stellar contribution of the Galactic disc can be splitted into two discs — a thin and a thick disc — which are usually parametrised by an exponentially decreasing density distribution. The parametrisation of the Galactic disc is taken from [48]

$$\rho_d(r,z) = \rho_d \cdot exp(-r/r_d) \cdot exp(-z/z_d). \tag{3.33}$$

The parameter ρ_d describes the density of the Galactic disc at the GC while r_d and z_d describe the scale parameter in radial and vertical direction. There is some freedom in the choice of the parameters for the Galactic disc. As in case of the bulge its density in the GC is unknown, so it has to be a free parameter. The scale radius is adopted from [122]

$$r_d = 2.3 \pm 0.6 \text{ kpc.} \tag{3.34}$$

The scale height z_d varies for the different components: young stars are born near $z = 0$, so they have a much smaller scale height, while older stars have been kicked around and reach scale heights up to 700 pc. Consequently, the disc has two distinct populations with two different scale heights. The thin disc consists of young, bright and metal-rich stars which provide about 98% of the total stellar population [123]. Its scale height was determined from a counting of stars with an absolute magnitude of $M_V > 6$ to be $z_d \approx 270$ pc [124]. From measurements of the spatial distribution of K dwarf stars the vertical scale height turns out to be smaller than 250 pc [125]. In [126] the scale height of the thin disc was determined for stars with $M_V \gtrsim 3.5$ to be 260 ± 50 pc. Furthermore the interstellar gas and dust contribution is also located in the thin disc. The thick disc consists of old, metal-poor stars and could be the result of either an earlier thin disc or the merging of a satellite galaxy with the MW [125]. Its scale height lies between 700 and 1500 pc and its local density is about 5% of the local stellar density of the thin disc. For this reason the luminous matter contribution is given only by the thin disc while small density contribution of the thick disc is neglected in this analysis. For gas, which makes up 10% of the mass of the disc, the scale height varies with Galactocentric radius because of the decreasing gravitational potential at larger radii. Fortunately, the mass model of the Galaxy is not very sensitive to the exact value of the scale height, so it is fixed to 320 pc, which is close to the value adopted by [127]. The hole in the centre of the disc [127], the bar structure and the black hole are not taken into account in the parametrisation, since the parameters of interest, i.e. the DM halo parameters, are insensitive to the detailed mass distribution in the centre of the Galaxy. The parametrisation of the visible mass discussed above leads to a mass of the Galactic bulge of about $1.1 \cdot 10^{10}$ $M_\odot$. The mass of the Galactic disc varies in the range of $5 \cdot 10^{10}$ to $7 \cdot 10^{10}$ solar masses for different fits because of the variation of the parameters ρ_d and r_d.

3.2.2 Dark Matter density

In addition to the baryonic matter the density distribution of the DM component has to be parametrised. The first analyses of the structure formation in the Universe in 1977 [128, 129] predicted that DM in Galactic haloes is distributed according to a simple power law distribution $\rho(r) \propto r^{-\gamma}$. However, later works based on numerical N-body simulations [76, 130, 131] found that the slope of the density distribution in the DM halo is different for different scales of the galactocentric distance. Today it is commonly believed that the DM distribution in a spherical halo can be well fitted by an universal function with four parameters. This parametrisation is given by

$$\rho_\chi(r) = \rho_{\odot,DM} \cdot \left(\frac{r}{r_\odot}\right)^{-\gamma} \left[\frac{1 + \left(\frac{r}{a}\right)^\alpha}{1 + \left(\frac{r_0}{a}\right)^\alpha}\right]^{\frac{\gamma-\beta}{\alpha}}, \tag{3.35}$$

where r is the distance to the Galactic centre, $r_\odot$ is the galactocentric distance to the Sun, a is the scale radius of the density profile and $\rho_\odot$ is the DM density of the halo at the position of the Sun. The parameters α, β and γ characterise the radial dependence of the density distribution. For $r \approx a$ the slope of the halo profile is about $r^{-\alpha}$, for $r \gg a$ holds a radial dependence of $r^{-\beta}$ and the slope of the halo profile in the centre of the Galaxy $r \ll a$ is about $r^{-\gamma}$. Many different sets for these parameters were suggested by different

Profile	α	β	γ	a [kpc]	Reference
NFW	1.0	3.0	1.0	20.0	[135]
BE	1.0	3.0	0.3	10.2	[132]
Moore	1.5	3.0	1.5	30.0	[76]
PISO	2.0	2.0	0.0	5.0	[16]
240	2.0	4.0	0.0	4.0	

Table 3.2: Parameter settings for the different DM halo profiles considered in this analysis. The parameters of the NFW profile are adapted from the publication by [135] but in this analysis a larger scale radius a is used.

analyses. In general, the results from numerical simulations show that the DM density in the Galactic centre is divergent. Such profiles are called cuspy profiles due to the cusp in the Galactic centre. This analysis considers several cuspy profiles. In 1997 Navarro, Frenk and White [130] discovered that their simulation results can be approximated by a profile with a slope of $\gamma = 1$ (hereafter referred to as NFW profile). Later simulations in 1999 by Moore et al. [76] prefer a profile with a steeper slope of $\gamma = 1.5$ (hereafter referred to as Moore profile). A third cuspy halo profile was introduced in [132]. Arguing that the microlensing data toward the Galactic centre produced by the MACHO collaboration is in conflict with a density profile with $\gamma \gtrsim 0.3$ (hereafter referred to as BE profile). In contrast to cuspy profiles the density distributions preferred by observations of rotation curves of low surface brightness and dwarf spiral galaxies have a nearly constant DM density in the Galactic centre ($\gamma \approx 0$) [128, 129, 133, 134]. Such profiles are called cored profiles due to the constant density in the central region. Two different cored halo profiles are considered in this analysis. The first profile is called pseudo-isothermal profile (hereafter PISO) since it is an isothermal profile ($\propto r^{-2}$) which is flatted in the centre. The second cored halo profile (hereafter 240) is similar to the PISO profile but decreases rather strongly for large radii. The parameter settings for these profiles are summarised in Table 3.2 and their radial dependence is shown in Figure 3.3.

3.3 Results

3.3.1 Local DM density

In the last section a parametrisation of the DM density distribution was introduced which includes five parameters — three slopes α, β and γ, one scale radius a and the local DM density $\rho_{\odot,\text{DM}}$. The local DM density is a priori unknown. However, a reliable determination of this value is of great interest for direct DM searches, where elastic collisions between the WIMPs and the target material of the detector are searched for [14]. This signal is

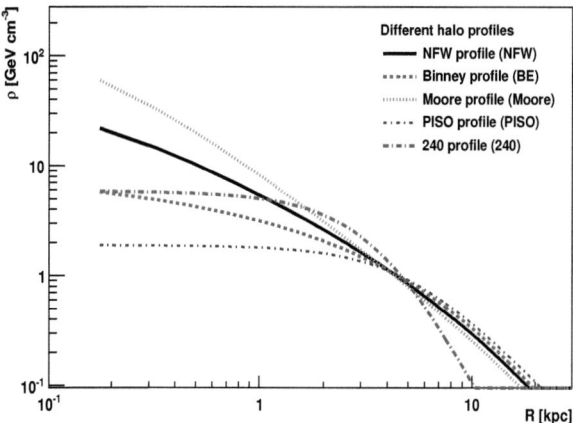

Figure 3.3: The radial dependence of the different halo profiles is shown under the requirement of equal mass inside the Solar orbit.

proportional to the local DM density. In the past the local DM density was determined within a large range — 0.2 to 0.7 GeV cm^{-3} [136, 137]. The upcoming of new astronomical data during the last years, discussed in Section 3.1, makes it worthwhile to reconsider the determination of this value.

The three most important constraints for the density model of the Galaxy are given by the rotation curve, the value $v_\odot$, the total mass M_{tot} and the local mass density $\rho_{\odot,tot}$. This can be easily seen as follows: $v_\odot^2 = v_{vis}^2 + v_{DM}^2$, where v_{vis}^2 and v_{DM}^2 are proportional to ρ_{vis} and ρ_{DM}, respectively. For a given halo profile M_{tot} is determined by ρ_{DM}, while the Oort limit $\rho_{\odot,tot}$ determines $\rho_{\odot,tot} = \rho_{vis} + \rho_{DM}$. So in principle one has 3 constraints with only 2 variables ρ_{vis} and ρ_{DM}, if the shapes of the DM halo and the visible matter were known.

Unfortunately, additional important parameters are i) the eccentricity of the DM halo ii) the concentration of the DM halo iii) the scale length of the disc and iv) the mass in the bar/bulge. In addition the mass model is sensitive to the geometry, i.e. the Galactocentric distance from the Sun $d_\odot$ and the halo profile. Additional constraints come from the surface density, but as discussed before here the visible surface density has a large uncertainty. The parameters and constraints have been summarised in Table 3.3. The parametrisation of the mass of the bulge was chosen to describe the rotation curve at small radii, which works reasonably well, as can be seen from Figure 3.5. Given that the mass model is not very sensitive to this inner region, the parameters of the bulge will not be varied anymore.

To optimise the remaining parameters in order to best describe the data, the following χ^2

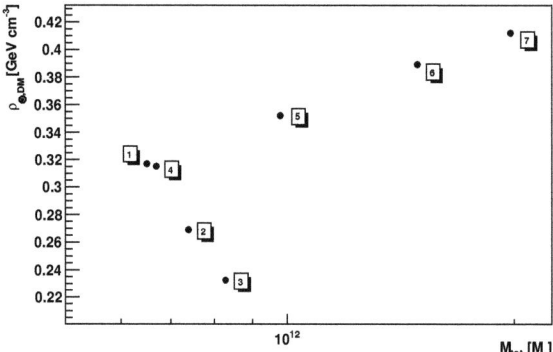

Figure 3.4: The local DM densities $\rho_{\odot,\mathrm{DM}}$ are shown for different fits with different parameters. The numbers correspond to the numbers of the fit results in Table 3.4. In 1 – 3 the scale parameter a is fixed and the scale radius of the Galactic disc r_d is left free. In 4 – 7 the scale radius of the disc r_d is fixed and a is left free.

function was minimised using the Minuit package [138]

$$\chi^2 = \frac{(M_{tot}^{calc} - M_{tot}^{exp})^2}{\sigma_{M_{tot}}^2} + \frac{(v_{\odot}^{calc} - v_{\odot}^{exp})^2}{\sigma_{v_{\odot}}^2} + \frac{(\rho_{tot}^{calc} - \rho_{tot}^{exp})^2}{\sigma_{\rho_{tot}}^2} + \frac{(\Sigma_{vis}^{calc} - \Sigma_{vis}^{exp})^2}{\sigma_{\Sigma_{vis}}^2} + \frac{(\Sigma_{tot}^{calc} - \Sigma_{tot}^{exp})^2}{\sigma_{\Sigma_{tot}}^2} + \frac{(r_d^{calc} - r_d^{exp})^2}{\sigma_{r_d}^2} + \frac{(Slope_{\odot,RC}^{calc} - Slope_{\odot,RC}^{exp})^2}{\sigma_{Slope_{\odot,RC}}^2} + \frac{((A-B)^{calc} - (A-B)^{exp})^2}{\sigma_{A-B}^2} \quad (3.36)$$

The index ^{calc} means the observables were calculated from the fitted parameters, while the index ^{exp} denotes the experimental data for the observable and σ its error. The constraints are summarised in Table 3.3.

The fit shows a more than 95% positive correlation between the local dark matter density and the scale length of DM halo a and an equally large negative correlation with the scale length r_d of the baryonic disc. Consequently, it is difficult to leave parameters free in the fit. Therefore the fit was first performed for fixed values of a (rows 1–3 of Table 3.4) and then r_d was fixed (rows 4–7). With the other free parameters all experimental constraints could be met within errors, as indicated by the χ^2 values in brackets below the fitted values in Table 3.4. Of course, the total mass changed for the different fits. Figure 3.4 shows the resulting local DM density versus the total mass, as calculated from the fitted parameters. It shows that in spite of the small errors for the local density for individual fits the spread in density

Free Parameters				
Component	Parameter	Symbol	Value	Unit
Halo	Local DM density	$\rho_{\odot,\text{DM}}$	-	GeV/cm^{-3}
Halo	Scale Parameter	a	-	kpc
Halo	eccentricity	ϵ_z	-	kpc
Disc	Density at GC	ρ_d	-	GeV cm^{-3}
Disc	scale length	r_d	-	kpc
Fixed Parameters				
Disc	scale height	z_d	320	pc
Bulge	Density at GC	ρ_b	360	GeV cm^{-3}
Bulge	Eccentricity	q_b	0.61	
Bulge	Scale	r_t	0.6	kpc
Bulge	Scale	$r_{0,b}$	0.7	kpc
Bulge	Slope	γ_b	1.6	
Bulge	Slope	β_b	1.6	
Constraints				
All	Mass inside 60 kpc	$M_{R<60\text{kpc}}$	4.0 ± 0.7	10^{11} M$_\odot$
Local	Rotation speed Sun	$v_\odot$	244 ± 10	km s^{-1}
Local	Distance Sun-GC	$r_\odot$	8.33 ± 0.35	kpc
Local	Total Surface Density	$\Sigma_{|z|<1.1}$	71 ± 6	M$_\odot$ pc^{-2}
Local	Visible Surface Density	Σ_{vis}	48 ± 9	M$_\odot$ pc^{-2}
Local	Mass Density	ρ_{tot}	0.102 ± 0.01	M$_\odot$ pc^{-3}
Local	Oort Constants	A-B	29.45 ± 0.15	km s^{-2} kpc^{-1}
Local	Slope of rotation curve	$\partial ln(v_\odot)/\partial ln(r)$	-0.006 ± 0.016	

Table 3.3: Free and fixed parameters for the density model of the Galaxy and experimental constraints. One observes that there are 6 free parameters and 8 constraints. Mass densities are in GeV cm^{-3} or in $M_\odot$ pc^{-3}, where 1 M$_\odot$ cm^{-3} ≡ 37.97 GeV cm^{-3}.

66 Constraint on the Dark Matter density distribution from astronomical observations

Nr	Profile	Fitted Parameters					Derived Quantities							
		$\rho_{\odot,DM}$	a	ρ_d	r_d	c_{vir}	$\rho_{\odot,tot}$	$v_\odot$	M_{60}	M_{tot}	A-B	$\frac{\partial \ln(v_\odot)}{\partial \ln(r)}$	$\Sigma_{|z|<1.1}$	Σ_{vis}	χ^2
1	NFW	0.32 ± 0.05	10	88.3 ± 19.8	2.5 ± 0.2 (0.13)	17.5	0.094 (0.58)	244.5 (0.0)	$3.9 \cdot 10^{11}$ (0.0)	$6.5 \cdot 10^{11}$ (0.01)	29.45 (0.01)	-0.002 (0.06)	72.2 (0.04)	53.8 (0.41)	(1.25)
2	NFW	0.27 ± 0.06	15	113.1 ± 17.1	2.4 ± 0.1 (0.01)	12.0	0.096 (0.38)	244.5 (0.0)	$4.0 \cdot 10^{11}$ (0.0)	$7.4 \cdot 10^{11}$ (0.0)	29.45 (0.0)	-0.003 (0.04)	71.1 (0.04)	55.4 (0.68)	(1.13)
3	NFW	0.23 ± 0.05	20	128.7 ± 69.8	2.3 ± 0.1 (0.0)	9.5	0.097 (0.27)	244.4 (0.0)	$4.1 \cdot 10^{11}$ (0.02)	$8.3 \cdot 10^{11}$ (0.0)	29.45 (0.0)	-0.003 (0.03)	70.2 (0.02)	56.6 (0.92)	(1.27)
4	NFW	0.32 ± 0.04	10.8 ± 3.4	91.0 ± 8.0	2.5	16.2	0.095 (0.50)	244.4 (0.0)	$4.0 \cdot 10^{11}$ (0.0)	$6.7 \cdot 10^{11}$ (0.0)	29.45 (0.0)	-0.003 (0.04)	72.4 (0.06)	54.2 (0.48)	(1.07)
5	NFW	0.35 ± 0.06	14.9 ± 4.8	89.5 ± 8.2	2.5	13.4	0.094 (0.57)	244.4 (0.0)	$4.25 \cdot 10^{11}$ (0.0)	$9.8 \cdot 10^{11}$ (0.0)	29.45 (0.0)	-0.003 (0.05)	73.7 (0.21)	53.2 (0.34)	(1.17)
6	NFW	0.39 ± 0.05	20.4 ± 4.5	88.1 ± 8.7	2.5	11.5	0.094 (0.62)	244.4 (0.0)	$5.5 \cdot 10^{11}$ (0.0)	$1.49 \cdot 10^{12}$ (0.0)	29.44 (0.0)	-0.001 (0.08)	75.0 (0.45)	52.4 (0.24)	(1.40)
7	NFW	0.41 ± 0.05	25.2 ± 4.6	87.3 ± 8.9	2.5	10.1	0.094 (0.64)	244.4 (0.0)	$6.5 \cdot 10^{11}$ (0.0)	$1.98 \cdot 10^{12}$ (0.0)	29.44 (0.0)	-0.001 (0.09)	75.9 (0.68)	51.9 (0.19)	(1.61)
8	BE	0.25 ± 0.05	10.2	133.6 ± 15.2	2.29 ± 0.09 (0.0)	17.6	0.096 (0.31)	244.4 (0.0)	$4.1 \cdot 10^{11}$ (0.01)	$7.5 \cdot 10^{11}$ (0.0)	29.45 (0.0)	-0.003 (0.03)	70.6 (0.01)	56.2 (0.84)	(1.20)
9	Moore	0.25 ± 0.05	30.0	114.7 ± 17.3	2.36 ± 0.11 (0.01)	6.2	0.095 (0.32)	244.4 (0.0)	$4.1 \cdot 10^{11}$ (0.01)	$7.6 \cdot 10^{11}$ (0.0)	29.45 (0.0)	-0.003 (0.04)	70.4 (0.01)	56.2 (0.82)	(1.21)
10	PISO	0.20 ± 0.04	5.0	150.4 ± 12.9	2.21 ± 0.08 (0.02)	46	0.098 (0.19)	244.4 (0.0)	$4.1 \cdot 10^{11}$ (0.03)	$1.45 \cdot 10^{12}$ (0.02)	29.45 (0.0)	-0.004 (0.02)	69.3 (0.08)	57.8 (1.18)	(1.54)
11	240	0.26 ± 0.03	4.0	53.1 ± 9.6	3.0 ± 0.2 (1.36)	26.3	0.095 (0.51)	244.6 (0.0)	$1.7 \cdot 10^{11}$ (10.45)	$1.8 \cdot 10^{11}$ (0.0)	29.47 (0.02)	-0.002 (0.06)	70.0 (0.03)	55.0 (0.61)	(13.04)
12	NFW	0.52 ± 0.07	26.6 ± 4.9	84.2 ± 9.5	2.5	9.6	0.094 (0.66)	244.3 (0.01)	$6.6 \cdot 10^{11}$ (0.0)	$1.95 \cdot 10^{12}$ (0.03)	29.43 (0.02)	-0.001 (0.10)	80.3 (2.38)	50.1 (0.66)	(3.24)

Table 3.4: Fit results. The units of the different values are given in Table 3.3.

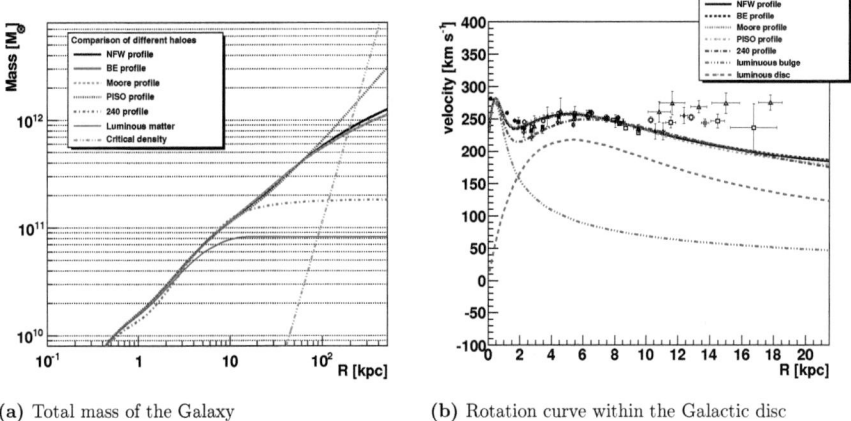

(a) Total mass of the Galaxy (b) Rotation curve within the Galactic disc

Figure 3.5: On the left side the mass inside a radius as a function of that radius is shown for the different halo profiles defined in Table 3.2. The thin solid line represents the visible mass which is different for different halo profiles because of the variation of the parameters ρ_d and r_d. Here the luminous matter for the NFW profile is shown. The steep line starting at 40 kpc represents the mass of a sphere with a density of 200 times the critical density of the Universe. The crossings of the mass distributions with this line represent the virial radius and the total Galactic mass of the corresponding density distribution. On the right side the rotation curve - calculated for different halo profiles - is compared with experimental data.

is still quite large.

The fit was repeated for other halo profiles yielding similar χ^2 values as shown in the rows 9-11 in Table 3.4. That means that with the present data one cannot distinguish the different halo profiles. So far only spherical halos have been discussed. Allowing oblate halos with a ratio of short-to-long axis of 0.7 the local DM density increases by about 20%, as shown by the last row of Table 3.4. An additional amount of DM in the Galactic disc corotating with the stellar matter, so-called dark disc [139], can enhance this value considerably more, so the uncertainty usually quoted for the local dark matter density in the range of 0.2 to 0.7 GeV cm^{-3} [136, 137] is still valid in spite of the considerably improved data.

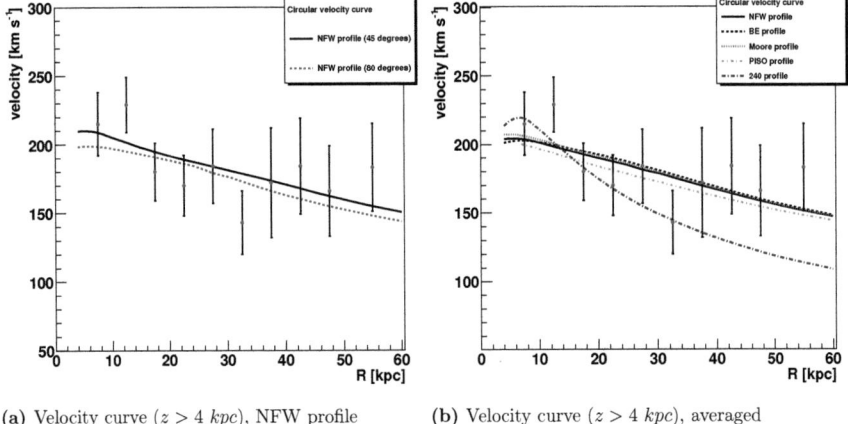

(a) Velocity curve ($z > 4\ kpc$), NFW profile (b) Velocity curve ($z > 4\ kpc$), averaged

Figure 3.6: The circular velocity curve for halo stars above a height of $z > 4$ kpc. On the left side the circular velocity curve for the NFW profile calculated for different angles with respect to the Galactic disc is shown. On the right side the averaged circular velocity curves for the different halo profiles are shown. The experimental data were obtained from the publication by [100].

3.3.2 Ringlike Dark Matter substructure

In the previous section it was shown that astronomical data can be well described by cuspy and cored profiles, simply because the gravitational potential in the GC is dominated by baryonic matter. However, the velocity distribution in the outer Galaxy is not compatible with a pure halo profile. In order to explain the increase of the rotation velocity at larger radii an additional ringlike substructure of the DM is needed. The parametrisation of the DM density distribution is modified to include a ringlike structure. In [92] such a structure is assumed to consist of wavy rings which are accompanied by a depletion at smaller radii. Instead this analysis considers the ring to be the result of the infall of a dwarf galaxy. In this case there is no depletion at smaller radii.

Disclike structures, which co-rotate with the luminous matter, are expected from N-body simulations of the accretion of satellite galaxies onto early galactic discs [140]. Coplanar tidal streams resulting from the disruption of the satellite galaxy only feel the radial gradient of the gravitational potential of the Galaxy, which leads to ringlike structures with a much longer lifetime than the tidal streams in the halo. The tidal streams in the halo are quickly destroyed by tidal shocks during the passage through the Galactic disc. Such a ringlike structure in the MW, the Monoceros ring, has been traced almost completely around the Galaxy (see Section 2.5). An enhancement of stars along this ring was discovered in the Canis Major constellation [141, 142] at Galactic longitudes around 240°. This overdensity was interpreted as a dwarf galaxy, called Canis Major Dwarf, which could be the progenitor of the tidal stream. The velocity dispersion of the Canis Major stars is very low which further confirms their common origin [143] and is not explainable with a warp of the Galactic disc [47].

As the RC at the inner Galaxy is independent on the DM halo profile in this section the rotation velocities for a Galactocentric distance greater than 3 kpc are considered. The parametrisation of the luminous matter was given in Section 3.2.1. The parameters of the Galactic bulge did not change while the parameters of the Galactic disc are assumed to be $\rho_d = 100$ GeV cm^{-3} and $r_d = 2.3$ kpc. This parametrisation yield a local surface density of the luminous matter of about 45 M$_\odot$ pc^{-2} which is in good agreement with experimental data (see Section 3.1.4). The parametrisation of the Galactic DM component including rings is given by

$$\rho_\chi(\vec{r}) = \rho_{\odot,\text{Halo}} \cdot \left(\frac{\tilde{r}}{r_\odot}\right)^{-\gamma} \cdot \left[\frac{1 + \left(\frac{\tilde{r}}{a}\right)^\alpha}{1 + \left(\frac{r_\odot}{a}\right)^\alpha}\right]^{\frac{\gamma-\beta}{\alpha}} + \sum_{n=1}^{2} \rho_n \exp\left(-\frac{(\tilde{r}_{gc,n} - R_n)^2}{2 \cdot \sigma_{R,n}^2} - \left|\frac{z}{\sigma_{z,n}}\right|\right), \quad (3.37)$$

$$\tilde{r} = \sqrt{x^2 + \frac{y^2}{\epsilon_{xy}^2} + \frac{z^2}{\epsilon_z^2}}, \qquad \tilde{r}_{gc,n} = \sqrt{x_{(n)}^2 + \frac{y_{(n)}^2}{\epsilon_{xy,n}}},$$

where the first term represents the Galactic halo as given in the previous section and the second term describes the DM rings. In this extended DM density distribution the halo is not spherically shaped anymore. The eccentricities ϵ_{xy} and ϵ_z describe the flattening of the halo profile within the Galactic disc and perpendicular to the disc. A further degree of freedom concerning a triaxial halo profile is the angle ϕ_{GC} between the major axis of the halo profile in the Galactic plane and the connection line between the Sun and the GC. In both rings the DM distribution in z-direction decreases exponentially with the scale heights $\sigma_{z,n}$ and Gaussian distributed in r-direction around the ring radius R_n and a width $\sigma_{R,n}$. The DM rings are allowed to be elliptical. Their eccentricities are given by the parameters $\epsilon_{xy,n}$. Like for the halo, the DM rings can be turned around the respective angles ϕ_n which is the angle between the major axis of the ring and the connection line Sun - GC. Here, a spherical NFW profile is assumed for the halo density distribution. Its scale radius is fixed to $a = 12$ kpc. The radial width of the outer ring is assumed to be different for the inner and the outer side. This difference results from the conservation of angular momentum. The remnants of an infalling dwarf galaxy with predetermined angular momentum with respect to the GC cannot reach arbitrary small Galactocentric distances. Therefore, a Gaussian distribution for the radial decrease of the inner side of the outer ring with a non-zero density at the GC is disfavoured for this reason. In this analysis the radial density decrease of the inner side of the outer ring is parametrised with two parabolic functions as an s-shape

$$\rho_{OR}(r) = \begin{cases} \frac{4\rho_{OR}}{d_{OR}^2} \cdot (r - (R_{OR} - d_{OR}))^2 & for \quad (R_{OR} - d_{OR}) < r < (R_{OR} - d_{OR}/2) \\ \rho_{OR} - \frac{4\rho_{OR}}{d_{OR}^2} \cdot (r - R_{OR})^2 & for \quad (R_{OR} - d_{OR}/2) < r < R_{OR} \end{cases} \quad (3.38)$$

where the parameter d_{OR} is the distance in which the DM density decreases to zero.

The χ^2 function in Eq. (3.36) is minimised using the DM density distribution in Eqs. (3.37) and (3.38). The ring parameters of the best fit to the astronomical data are given in Table 3.3.2. These parameters provide DM rings with masses of $7.5 \cdot 10^9$ solar masses for the inner ring and $4.1 \cdot 10^{10}$ solar masses for the outer ring. The local halo density $\rho_{\odot,\text{Halo}}$ is found to be 0.31 GeV cm^{-3} while the total DM density at the position of the Sun is $\rho_{\odot,\text{DM}} = \rho_{\odot,\text{Halo}} + \rho_{\odot,\text{IR}} + \rho_{\odot,\text{OR}} = 1.0$ GeV cm^{-3}. The total DM density within the Galactic disc is shown in Figure 3.7a. There the influence of the DM rings is clearly visible. At the position of the inner ring the DM density is increased by about a factor of 5, while at

Parameter	Inner Ring	Outer Ring
$\rho_{R_{\text{ring}}}$ [GeV cm^{-3}]	3.3	1.8
r [kpc]	4.5	12.5
σ_r [kpc]	1.8	3.7
σ_z [pc]	330	660
d [kpc]	-	6.0
ϵ	1.0	1.0
ϕ [°]	0.0	0.0

Table 3.5: Ring parameters of the best χ^2 fit.

the outer ring the density is about 14 times as high as the density of the DM halo at this position. Although introducing DM rings increases the local DM density by a factor of three the total surface density is still compatible with experimental data. For the ring parameters in Table 3.3.2 a surface density of 78 $M_\odot$ pc^{-2} is obtained which is in good agreement with determinations of the local surface density described in Section 3.1.4. The total mass distribution of the Galaxy is presented in Figure 3.7b. In addition to the local surface density the mass constraint in Section 3.1.3 is fulfilled by this density distribution too. This compatibility results from the fact that the total mass of the Galaxy is provided by the DM halo. The DM rings only provide a few percent of the total Galactic mass. Therefore, the local halo density $\rho_{\odot,\text{Halo}}$ is almost only constrained by the Galactic mass (if the concentration or the scale radius a of the halo profile respectively is fixed). The velocity distribution in the Galactic disc is shown in Figure 3.7c. It is shown there that a spherical NFW profile in combination with two concentric DM rings at a Galactocentric distance of 4.5 and 12.5 kpc yield a good description of the change of slope at 9 kpc. The circular velocity curve of halo stars in Figure 3.7d is in agreement with the experimental data, although halo stars in a Galactocentric distance of about 20 kpc are measured with lower rotation velocities than the rotation velocities predicted by the density model. The estimation of the HWHM of the Galactic gas component from such an extended DM halo is shown in Figure 3.8a. The description of the HWHM is improved by the introduction of a DM substructure, even though the estimated gas flaring is too low in the radial region around 15 kpc. A lighter ring with a mass of about $2.0 \cdot 10^{10}$ solar masses, as found in [20], yields a good fit of the HWHM, but is incompatible with the outer RC, as shown in Figure 3.8b.

Nevertheless, a DM halo profile with ringlike substructure composed of two DM rings is compatible with astronomical observations of the RC in the Galactic disc, the circular rotation curve of halo stars, the local surface density and the total matter density at the Sun.

3.4 Discussion

In this section it is shown that the astronomical constraints are consistent with a density model of the MW consisting of a central Galactic bulge, a Galactic disc and an extended DM halo. No sensitivity of the astronomical constraints concerning the discrimination between cuspy and cored profiles is found since the total matter density in the GC is dominated by the visible matter of the MW. The local DM density is obtained to be in the range between 0.2 and 0.4 GeV cm^{-3}. Strong positive and negative correlations between the considered parameters were found from the fit and they are causing the obvious correlation between $\rho_{\odot,\text{DM}}$ and M_{tot}. For non-spherical haloes these values can be enhanced by 20%. If dark discs are considered [139], densities up to 0.7 GeV cm^{-3} can easily be imagined, so the previous range of 0.2 to 0.7 GeV cm^{-3} still seems valid. The velocity distribution in the Galactic disc shows a change of slope which cannot be described by the considered halo profiles.

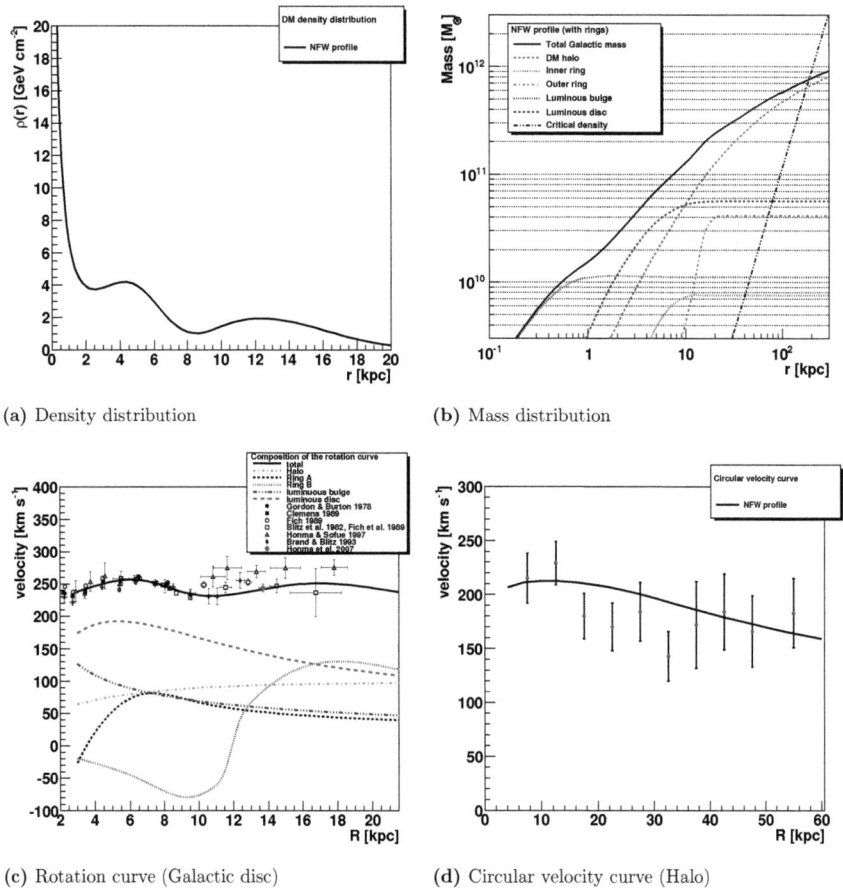

Figure 3.7: Results of an DM halo profile combined with a DM substructure of two circular DM rings at 4.5 and 12.5 kpc. (a) The radial dependence of the DM density is shown. The influence of the DM rings is clearly visible. (b) Mass distribution of an NFW profile with substructure. The crossing with the straight line represents the virial mass of the Galaxy and its virial radius. (c) The RC in the Galactic disc for an NFW halo profile in combination with two concentric DM rings. (d) The circular velocity curve of halo stars, averaged from estimations of velocity curves at Galactic latitudes of 45° and 80°.

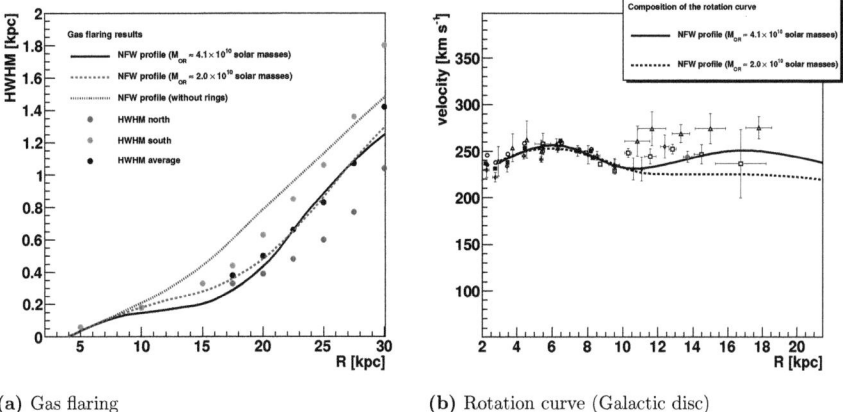

(a) Gas flaring (b) Rotation curve (Galactic disc)

Figure 3.8: Gas flaring estimation of an DM halo profile combined with a substructure of two circular DM rings at 4.5 and 12.5 kpc. **(a)** HWHM of the Galactic gas distribution estimated with an NFW profile combined with two DM rings. For comparison the HWHM estimated with an NFW profile without rings is shown. **(b)** The RC within the Galactic disc. The increase of the rotation velocity at about 10 kpc can be described by a DM ring outside the solar circle.

It is shown that a halo profile in combination with a DM substructure composed of two concentric rings in the Galactic disc describes the RC. The Galactocentric distance of the rings is obtained to be 4.5 and 12.5 kpc. Although the DM density is increased by a factor of five at the position of the inner ring and a factor of 14 at the position of the outer ring the astronomical observations can be met because the Sun is located between the rings where the DM density is low. The DM rings are assumed to be the result of the infall of a dwarf galaxy in the gravitational potential. Numerical simulations show that ringlike structures in galaxies like the MW can be produced in such a scenario [144].

4
Constraints on the Dark Matter density distribution from Galactic gamma rays

In the previous chapter the Galactic DM density distribution and especially the local DM content were considered with respect to astronomical observations of the MW. In this chapter the annihilation of DM particles is taken into account. Assuming that WIMPs annihilate each other into charged particles and gamma rays the spatial distribution of the resulting gamma radiation can be used to estimate the density distribution within the DM halo. Charged annihilation products cannot be used for this purpose since they are strongly influenced by the Galactic magnetic field while the gamma rays directly point back to the place of their production. The gamma radiation produced by DMA consequently contributes to the diffuse Galactic gamma radiation which was measured with the Energetic Gamma Ray Emission Telescope (EGRET) experiment. The photon energy spectrum measured by this experiment shows an excess above a photon energy of about 1 GeV which can be interpreted as an gamma ray signal from DMA. A method for the reconstruction of the DM density distribution using the diffuse Galactic gamma radiation is presented in the next section. Subsequently, the diffuse Galactic gamma radiation measured with EGRET, which shows an excess above a photon energy of about 1 GeV, is analysed according to this procedure. In addition, new preliminary data from the Fermi satellite, the successor of EGRET, is analysed. At the end of the chapter recent observations of the diffuse Galactic gamma radiation are discussed and upcoming experiments are explained.

4.0.1 The EGRET experiment

The EGRET [145–147] experiment was one of four telescopes (BATSE, OSSE, COMPTEL and EGRET) on the Compton Gamma Ray Observatory (CGRO) satellite which operated

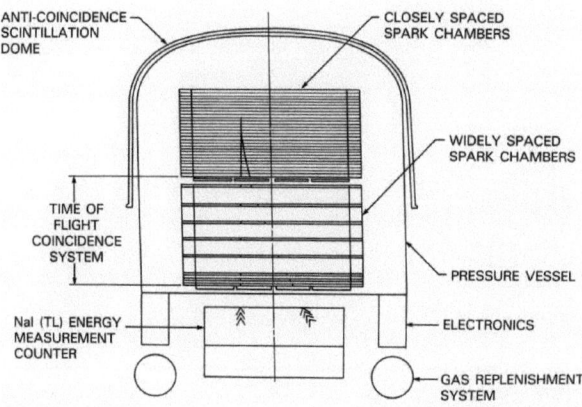

Figure 4.1: Schematic diagram of the EGRET instrument. The anticoincidence dome is shown at the top. The direction and the energy of a photon, which produces an electron-positron pair in the spark chamber, can be reconstructed from the traces of the leptons in the spark chamber and their energy deposit in the calorimeter. Figure taken from [75].

in the orbit of the Earth for nearly 10 years (1991-2000). During this time CGRO measured the diffuse gamma radiation from energies of 15 keV to energies above 50 GeV. Compared to all other experiments on the CGRO satellite EGRET covered the highest energy range from approximately 30 MeV to 30 GeV. The recorded data are publicly available on the NASA's webpage [148]. At the high end of the energy range the sensitivity was limited is low counting statistics. EGRET was calibrated at the Stanford Linear Accelerator Center (SLAC) and tested for proton-induced background at Brookhaven. A schematic picture of the EGRET instument is shown in Figure 4.1. The central element of EGRET was a multilevel spark chamber which was triggered by a directional scintillator coincidence system. A Total Absorption Shower Counter (TASC) which consisted of 36 NaI(Tl) blocks was installed below the spark chamber in order to measure the event energy. All calorimeter blocks were optically coupled to form a monolithic scintillator which was viewed by 16 photomultiplier tubes (PMT). The instrument was covered by a scintillator dome which was used in anticoincidence with the trigger system to veto charged particles. The gamma rays entering the scintillator dome produced electron-positron pairs in the tantalum foils which were located between the spark chamber tracking layers. The trigger coincidence system of the spark chamber was installed between the lowermost tantalum foil and the bottom of the tracker. It consisted of two 4×4 arrays of plastic scintillator tiles (one at the top and one

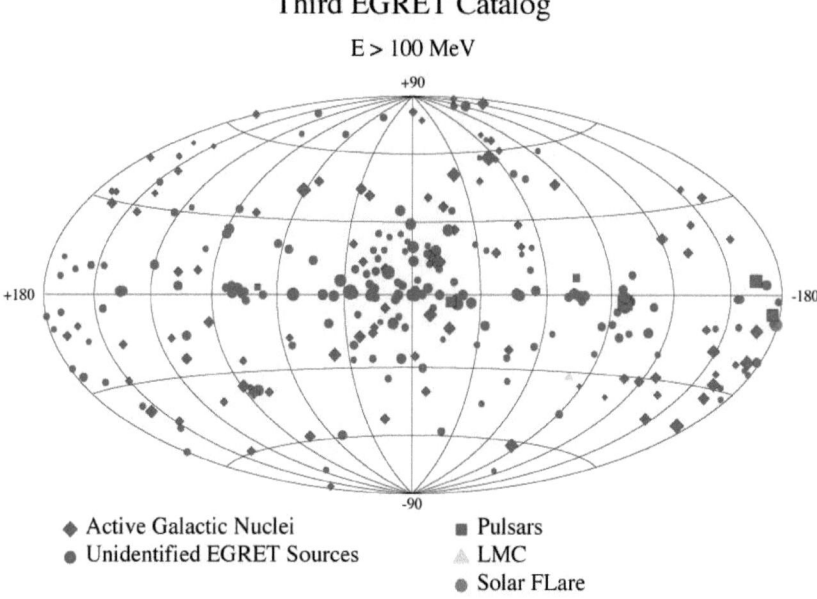

Figure 4.2: The third EGRET catalogue is one of the main results of the experiment. The main contribution of the resolved sources are the active galactic nuclei (AGN) which are extragalactic sources and unidentified sources. Galactic sources, the Sun and pulsars, are located at small latitudes in the Galactic disc. The Magelanic Clouds, our neighbour galaxies, are also well resolved in this catalogue. Taken from [149].

at the bottom of the chamber). Furthermore, a time-of-flight (TOF) trigger system between these two scintillator planes was installed in order to indicate downward-moving particles. Therefore, an event trigger was produced by the measurement of the directional coincidence of a signal produced by an electron or a positron, together with a signal of the TOF.

As the EGRET instrument was originally designed for 2 years it was a great success that its operation time was nearly 10 years, although most data on diffuse gamma rays were collected between 1991 and 1994. During that time EGRET measured not only diffuse gamma radiation but also solar flares, pulsars, active galactice nuclei and other point sources. These pointlike sources are summarised in the *Third EGRET Catalogue* which is shown in Figure 4.2.

4.1 Reconstruction of the Dark Matter density distribution

The reconstruction method of the Galactic DM density distribution is based on the simultaneous consideration of

1. the energy spectrum of the diffuse Galactic gamma radiation, and
2. the directional distribution of the gamma radiation fluxes.

These distributions are sensitive to two different properties of the gamma radiation from DMA. The photon energy spectrum depends on the mass of the WIMP while the spatial distribution of the gamma radiation indicates the DM density distribution in the Galactic halo. In [16] this method was used to analyse the diffuse Galactic gamma radiation assuming a smooth distributed DM component. Here, this analysis is reconsidered assuming a DM contribution composed a smooth distributed and a clumpy distributed DM component. Next the photon energy spectrum and the spatial distribution of the gamma ray fluxes are explained in more detail.

A special feature of the presented method is that only the shapes of the different contributions (background radiation and DMA signal) are scaled in order to describe the energy spectrum of the diffuse Galactic gamma radiation. Such a data-driven calibration of the intensity of signal and background is widely used in experimental particle physics in order to cancel out large, correlated and model-dependent systematical errors. Then the uncorrelated point-to-point error, which can be interpreted as the error of the spectral shape, remains. The background radiation is composed of the gamma radiation from neutral pion decay, bremsstrahlung and inverse Compton scattering processes, as explained in Section 2.7. The dominant background contribution (neutral pion decay) results from the interaction of CR protons with interstellar gas. Its shape is known from fixed target experiments. As the CR proton suffer little energy losses during their propagation through the Galaxy the shape of this contribution is the same in all directions. The contributions of the CR electrons (bremsstrahlung and inverse Compton scattering) can offer little differences for different directions. The shape of the gamma ray signal from DMA is known from collision experiments with electrons and positrons. In this analysis the common shape of the background radiation, estimated from `GalProp` [79], is fitted to the EGRET data. Thereby, the scaling factor of the gamma ray signal from DMA (boost factor) is taken to be the same in all sky directions, i.e. that the averaged luminosity of the DMCs is equal in all directions. The scaling factors of the background radiation depend on the density distribution of CR and interstellar gas in the Galaxy, which are not well known. Therefore, the scaling factors of the background are left free for different sky regions. A fit of the shape of the individual background contributions cannot explain the EGRET gamma ray excess.

In a first step the spectral shape of the background model in the separate sky regions is

4.1 Reconstruction of the Dark Matter density distribution

fitted to the experimental data. Therefore, the χ^2 function

$$\chi^2 = \sum_i \left(\frac{f \cdot \phi^i_{bg} + \phi^i_{eg} - \phi^i_{exp}}{\sigma^i} \right)^2 \qquad (4.1)$$

is minimised. There, ϕ_{bg} represents the Galactic background radiation flux, ϕ_{eg} is the flux of the extragalactic background radiation, ϕ_{exp} is the measured diffuse Galactic gamma ray flux and σ is the point-to-point error. The scaling parameter of the background radiation flux is f while the index i denotes the specific energy bin. The second step is the approximation of an extended theoretical model which includes the gamma ray signal from the DM contribution. Then the χ^2 function changes to

$$\chi^2 = \sum_i \left(\frac{f \cdot \phi^i_{bg} + \phi^i_{DM,diff} + \phi^i_{DM,clump} + \phi^i_{eg} - \phi^i_{exp}}{\sigma^i} \right)^2. \qquad (4.2)$$

This function includes the annihilation flux from the smoothly distributed DM $\phi_{DM,diff}$ and the DMCs $\phi_{DM,clump}$, as given by Eqs. (2.55) and (2.61), respectively. For the determination of the DM density distribution the directionality of the gamma ray flux is considered. The gamma radiation flux predicted by the DM density distribution and the gamma ray fluxes from the Galactic and extragalactic background radiation are fitted to the experimental data in different sky regions. For this purpose a finer division of the sky is necessary in order to get proper information about the directionality of the gamma ray fluxes. Then a χ^2 function for the fit to the experimental fluxes is used again

$$\chi^2 = \sum_{i,j} \left(\frac{f_{i,j} \cdot \phi^{i,j}_{bg} + \phi^{i,j}_{DM,diff} + \phi^{i,j}_{DM,clump} + \phi^{i,j}_{eg} - \phi^{i,j}_{exp}}{\sigma^{i,j}} \right)^2. \qquad (4.3)$$

The various bins in latitudinal and longitudinal direction are characterised by the indices i and j. The background model is differently scaled in all $i \times j$ sky regions. For the determination of the background scaling factors the gamma ray fluxes are splitted into two energy ranges — a low energy range for photons below an energy of 500 MeV and a high energy range for photons above 500 MeV. At low energies the total gamma radiation flux is dominated by the background component which is the reason why the background scaling factors $f_{i,j}$ are obtained from a fit to the data in this energy range. The fit of the EGRET data shows a good agreement of the background scaling factors with the predictions of `GalProp`. The gamma ray fluxes in the high energy region are used for the determination of the Galactic DM density distribution. If the background components in all directions are known from the low energy region the additional gamma radiation fluxes $\phi^{i,j}_{DM,diff} + \phi^{i,j}_{DM,clump}$ can be fitted to the experimental data. The gamma radiation from the diffuse DM component is proportional to the line-of-sight integral of the squared DM density while the gamma ray flux from the clumpy DM component is only proportional to the line-of-sight integral of the DM density to the first power, as explained in Section 2.6.3. However, the total gamma ray

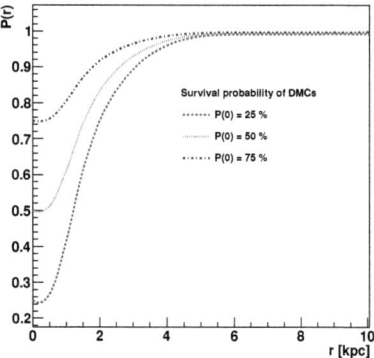

Figure 4.3: Radial dependence of the survival probability of DMCs in the MW. Different central values are obtained by analytical calculations depending on the ratio of the core radius and the virial radius of the DM clump, as explained in Section 2.6.3.

flux from DMA is dominated by the gamma ray flux from the clumpy component because of the increased annihilation rate of WIMPs in the core of DMCs which is reflected in the boost factor, which is found to be much larger then unity. Consequently, the analysis of the diffuse Galactic gamma is sensitive to the distribution of the clump DM component while the diffuse DM mass is mainly determined by astronomical observations.

The gamma ray flux from DMA in clumps also depends on tidal disruption of DMCs in the gravitational potential of the MW. This effect was discussed in Section 2.6.3. The survival probability of clumps in the Galaxy can be different from unity in the GC. In this thesis the survival probability P(r) is parametrised as [150]

$$P(r) = \frac{2.3 \cdot 10^6 - 1.0 \cdot 10^6 \cdot r + 1.7 \cdot 10^5 \cdot r^2 - 6.3 \cdot 10^3 \cdot r^3 + 91.1 \cdot r^4}{1.3 \cdot 10^5 - 1.7 \cdot 10^4 \cdot r + 2.2 \cdot 10^3 \cdot r^2 - 84.0 \cdot r^3 + 1.2 \cdot r^4} \cdot \exp(-(4/r)^4). \quad (4.4)$$

The maximal value of this function is normalised to unity, the radius is scaled by a factor of eight in order to obtain a saturation at a Galactocentric distance of about 5 kpc as indicated in [10]. The radial dependence of the survival probability is shown in Figure 4.3. In the following sections the diffuse Galactic gamma radiation measured with EGRET is analysed according the abovementioned procedure. The results in [16] are used as an indication for the parameters of the ringlike DM substructure.

4.2 Results

4.2.1 Energy spectrum of the diffuse Galactic gamma radiation

The energy spectrum of the diffuse Galactic gamma rays was found to be incompatible with expectations from a conventional and an optimised model of the Galactic gamma radiation [16]. It was shown that the fit of the energy spectrum can be much improved by the introduction of a gamma ray signal produced by annihilation of a WIMP with a mass above 50 GeV. The upper limit for the WIMP mass was found to be 70 GeV for a conventional model and 100 GeV for an optimised model. This result is still valid in case of mix of a smoothly and a clumpy distributed Galactic DM component since the energy spectrum is independent of the spatial distribution of the gamma fluxes. In the present thesis a WIMP mass of about 60 GeV is assumed which is in the abovementioned range.

Following the analysis in [16] the energy spectrum is considered in six different sky regions defined in Table 4.2.1. Region A is the energy spectrum in direction of the GC, where the DM contribution is assumed to be largest, while the energy spectrum of the gamma radiation from the Galactic disc without the GC is shown in region B. The spectrum of the Galactic anticentre (opposite to the direction to the GC) is given in region C. The last regions D, E, F are the spectra above the Galactic disc where the Galactic contributions are small and the extragalactic background radiation becomes important.

The χ^2 functions in Eq. (4.2) and Eq. (4.3) were minimised for the halo profiles defined in Section 3.2.2 in combination with two DM rings. In each case the extragalactic background radiation is calculated according to the method in [81]. Therein the extragalactic background is recursively calculated for each energy bin using the diffuse Galactic gamma

Region	Longitudes	Latitudes
A	$|l| < 30°$	$|b| < 5°$
B	$|l| > 30°$	$|b| < 5°$
C	$|l| > 90°$	$|b| < 10°$
D	$|l| < 180°$	$10° < |b| < 20°$
E	$|l| < 180°$	$20° < |b| < 60°$
F	$|l| < 180°$	$60° < |b| < 90°$

Table 4.1: Here the definition of the different sky regions for the analysis of the energy spectrum of the diffuse Galactic gamma radiation is shown. The coordinates l and b are given in the coordinate system centred at the Sun where b = 0° defines the Galactic disc and l = 0° defines the connection line between the Sun and the Galactic centre. More detailed information about the coordinate system in section 2.5.

radiation measured with EGRET. In this method the gamma ray flux produced by DMA is taken into account. In Figure 4.4 the energy spectra of the diffuse gamma radiation for the six different sky regions measured with EGRET are shown. The point-to-point error of the data, which represents the error of the spectral shape, is 7%. In case of an additional gamma ray flux from DMA with a WIMP mass of about 60 GeV the spectral fit is improved from $\chi^2/d.o.f. = 603.4/42$ to $\chi^2/d.o.f. = 30.9/36$.

4.2.2 Spatial distribution of the diffuse Galactic gamma radiation

In this section the results of the analysis of the spatial distribution of the diffuse Galactic gamma radiation are presented. Therefore, the sky is divided into 180 angular bins. The longitudes are divided into 45 bins with a width of 8° while the absolute values of the latitudes are divided into four bins: $0° < |b| < 5°$, $5° < |b| < 10°$, $10° < |b| < 20°$ and $20° < |b| < 90°$. The gamma radiation flux from the diffuse and the clumpy DM component is fitted to the experimental data according to Eq. (4.3). The density distributions of the diffuse and the clumpy DM component are treated separately which leads to two possible cases for the construction of the total DM density. In the first case one density distribution is assumed for both DM components. Hereafter, this model is called Single Profile (SP) density model. It corresponds to the analysis in Section 3.3.1 and 3.3.2. Then, the DM density is given by

$$\rho_{\chi,\text{tot}} = \rho_\chi^{\text{Halo}} + \rho_\chi^{\text{IR}} + \rho_\chi^{\text{OR}}, \qquad (4.5)$$

where the indices IR and OR denote the inner and outer ring. The shape of the DM halo profile is fixed and its normalisation $\rho_{\odot,\text{Halo}}$ is calculated under the requirement of the fulfillment of the local rotation velocity $v_\odot$.

In the second case the density distributions of the DM components are assumed to be different. Such density models are preferred by recent numerical simulations of the structure formation in the Universe [13]. Hereafter, this model is called Double Profile (DP) density model. Then the total DM density distribution is

$$\rho_{\chi,\text{tot}} = \rho_\chi^{\text{Halo,diff}} + \rho_\chi^{\text{Halo,clump}} + \rho_\chi^{\text{IR}} + \rho_\chi^{\text{OR}}. \qquad (4.6)$$

This model provides two local halo densities $\rho_{\odot,\text{Halo,diff}}$ and $\rho_{\odot,\text{Halo,clump}}$ which provides a further free parameter compared to the SP model. For the determination of the normalisation of the halo profile of the diffuse component the value of the total Galactic mass was used. It is known from experimental measurements of the kinematics of halo stars that the total Galactic mass of the MW is about $1.0 \cdot 10^{12}$ M$_\odot$. In order to obtain comparable results for the different profile combinations the diffuse halo was normalised in order to obtain a total mass for the diffuse component of $1.0 \cdot 10^{12}$ M$_\odot$ at a Galactocentric distance of 200 kpc. The local density of the diffuse component is consequently fixed if the parameters of the

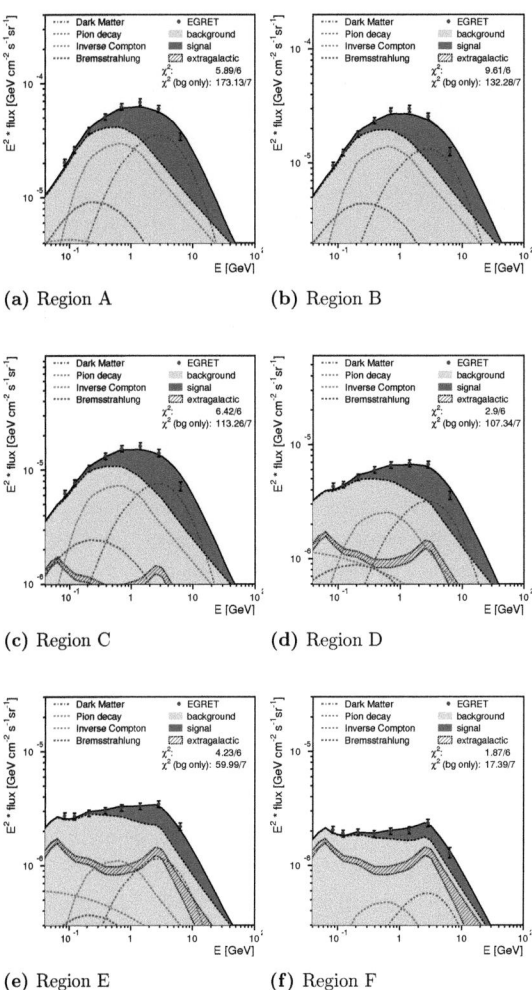

Figure 4.4: Energy spectrum of the diffuse Galactic gamma radiation in the six different regions defined in Table 4.2.1. The uncertainty of the data is 7%.

diffuse halo profile are fixed. The local density of the halo profile of the clumpy component is determined under the requirement of the fulfillment of the local rotation velocity $v_\odot$.
The diffuse DM distribution $\rho_\chi^{\text{Halo,diff}}$ is realised by a DM halo profile without rings. It describes the distribution of individual DM particles in the Galactic halo. The spatial distribution of DMCs and any other local enhancement of DM particles is given by $\rho_\chi^{\text{Halo,clump}} + \rho_\chi^{\text{IR}} + \rho_\chi^{\text{OR}}$. The DM rings are supposed to consist of tidal streams produced by the tidal disruption of DM clumps, the Sagittarius and the Canis Major Dwarf galaxy. Galactic tidal streams are supposed to belong to the clumpy DM component since they are small overdensities of DM. For this reason the clumpy DM component is assumed to consist of a halo profile in combination with the DM rings while for the smooth component a halo without rings is assumed.

Single Profile density model

In this section the results for the SP density model are presented. The minimisation of the χ^2 function in Eq. (4.3) was performed for the different halo profiles described in Section 3.2.2 combined with two DM rings. The scale radii a and the slopes α, β, γ of the different profiles were fixed. The halo eccentricities (ϵ_{xy} and ϵ_z) as well as the ring eccentricities ϵ_{IR} and ϵ_{OR} were adopted from [16]. The Galactocentric distance was fixed to $r_\odot = 8.3$ kpc. Since the rings are just slightly elliptical a small change of these values is hardly reflected in the fit results. The survival probability of the DMCs is assumed to be constant. In this case the gamma ray signal is dominated by clumps with a core radius smaller than 5% of the entire dimension of the clump (see Setion 2.6.3).
The parameter settings of the best fit to the spatial distribution of the gamma radiation flux, the χ^2 values and the boost factors are shown in Table 4.2. Comparing the fit results for the different density distributions it becomes apparant that the 240 profile yields the best description of the spatial distribution of the gamma ray fluxes. In case of the other halo profiles too small gamma ray fluxes are predicted for intermediate Galactic latitudes ($10° < |b| < 20°$) in the region of the GC. The reason for this is the radial dependence of the 240 profile, as shown in Figure 3.3, which provides a higher DM density in the radial region between 1 and 5 kpc compared to the other profiles. Consequently this profile provides a higher gamma ray flux in the intermediate latitude region. In order to illustrate this effect the longitudinal distribution of the NFW profile and the 240 profile are shown in Figures 4.5 and 4.6. The longitudinal gamma ray distributions of the remaining halo profiles are shown in Appendix C. All density distributions are dominated by the DM rings as shown in Figure 4.7a. This effect results from the inconsistency of a pure halo profile with the experimental data which was already found in [16], in which the DMA signal is only produced by the diffuse DM component. On the contrary, in the present thesis the major part of the DMA signal is produced by the clumpy component which is proportional to the line-of-sight integral of the linear DM density distribution. The linear dependence leads to even higher densities

	NFW	BE	Moore	PISO	240
$\rho_{\odot,\text{DM}}$ [GeV cm^{-3}]	0.128	0.064	0.169	0.062	0.131
$r_\odot$ [kpc]	8.3	8.3	8.3	8.3	8.3
α	1.0	1.0	1.5	2.0	2.0
β	3.0	3.0	3.0	2.0	4.0
γ	1.0	0.3	1.5	0.0	0.0
a [kpc]	20.0	10.2	30.0	5.0	4.0
ϵ_{xy}	0.85	0.85	0.85	0.85	0.85
ϵ_z	0.7	0.7	0.7	0.7	0.7
ϕ_{gc} [°]	87.0	85.3	83.4	83.3	83.0
ρ_{IR} [GeV cm^{-3}]	12.6	13.3	13.1	13.2	10.6
R_{IR} [kpc]	3.30	3.07	3.48	3.98	4.0
$\sigma_{r,\text{IR}}$ [kpc]	4.15	3.74	3.7	3.73	3.74
d_{IR}	-	-	-	-	-
$\sigma_{z,\text{IR}}$ [kpc]	0.323	0.366	0.306	0.317	0.29
$\epsilon_{xy,\text{IR}}$	0.85	0.85	0.85	0.85	0.85
ϕ_{IR} [°]	-76.3	-77.9	-82.8	-77.5	-73.1
M_{IR} [10^9 M$_\odot$]	45.9	45.1	30.7	47.2	34.6
ρ_{OR} [GeV cm^{-3}]	2.4	3.1	2.61	2.66	2.46
R_{OR} [kpc]	12.69	12.94	12.60	12.75	12.86
$\sigma_{r,\text{OR}}$ [kpc]	5.39	4.10	5.41	4.42	4.4
d_{OR}	4.0	4.0	4.0	4.0	4.0
$\sigma_{z,\text{OR}}$ [kpc]	1.0	1.07	1.06	1.02	1.02
$\epsilon_{xy,\text{OR}}$	0.95	0.95	0.95	0.95	0.95
ϕ_{OR} [°]	-50.7	116.4	121.2	126.6	48.0
M_{OR} [10^9 M$_\odot$]	10.5	11.4	12.1	9.7	9.0
boost factor	37.7	38.4	37.9	36.3	43.6
χ^2_{long} / d.o.f.	165.1 / 157	157.4 / 157	148.59 / 157	181.43 / 157	130.56 / 157
probability [%]	27.1	43.3	65.1	7.9	93.2

Table 4.2: Parameter settings of the SP model for the best approximations of the various DM halo profiles. The radius of the solar orbit r_0, the halo and ring eccentricities ϵ_{xy}, ϵ_z, ϵ_{IR} and ϵ_{OR} as well as the halo parameters a, α, β and γ were fixed. The initial ring parameters for the minimisation were adopted from [16].

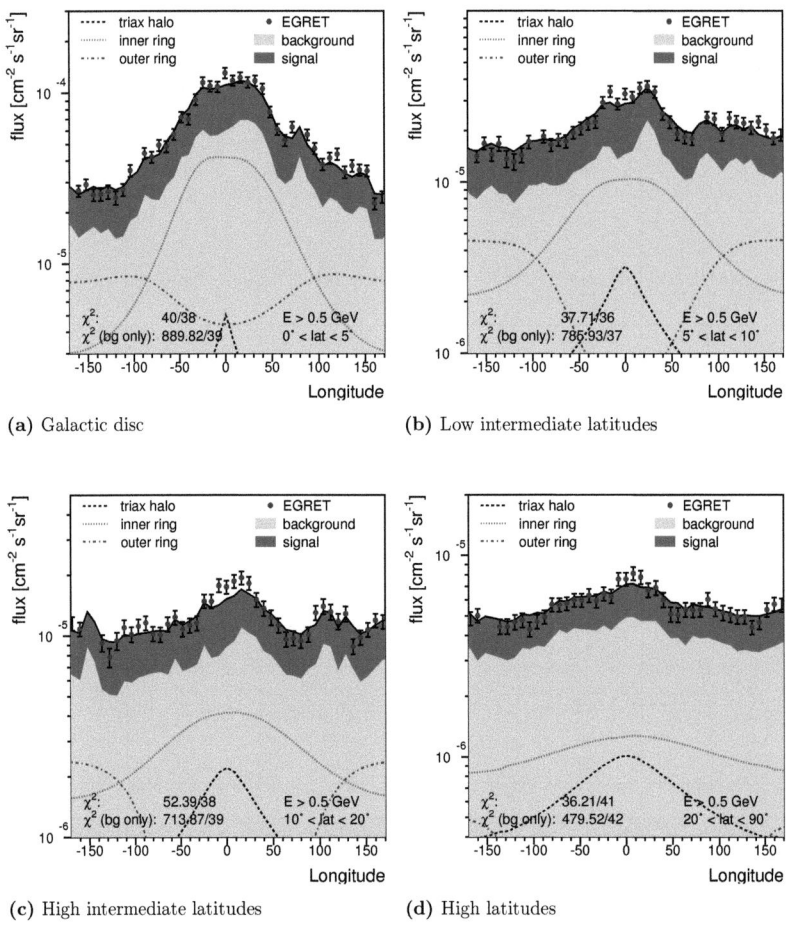

Figure 4.5: Longitudinal distribution of the diffuse Galactic gamma radiation obtained from the NFW profile in combination with a large scale structure of two rings at 3.3 and 12.7 kpc. A good fit of the data is obtained for latitudes up to 10° and at the Galactic poles. In the intermediate latitudinal region between 10° and 20° the intensity of the gamma radiation from the GC cannot be described.

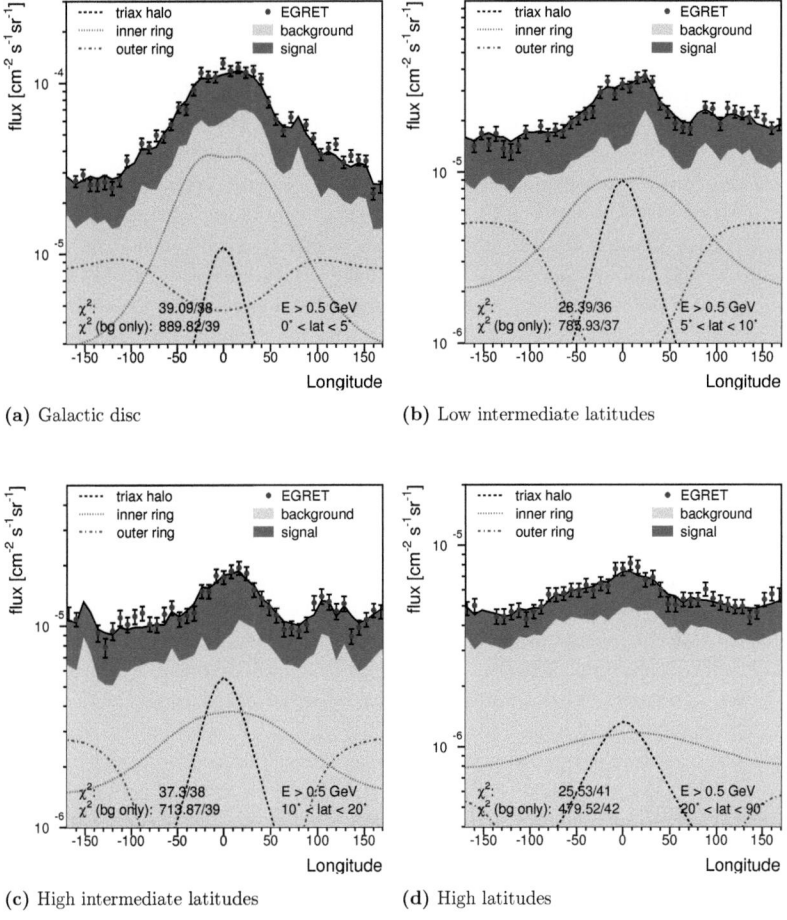

Figure 4.6: Longitudinal distribution of the diffuse Galactic gamma radiation obtained from the 240 profile in combination with a large scale structure of two rings at 4.0 and 12.9 kpc. A good fit of the data is obtained for latitudes up to 10° and at the Galactic poles. Contrary to the fits of the remaining profile in this case a good fit to the gamma radiation from the GC is obtained in the latitudinal region from between 10° and 20°.

compared to the case of just the diffuse DM component. Approximately equal densities are obtained for the outer ring for all profile, but a difference for the inner ring becomes apparent. Equal densities are obtained for the NFW, BE, Moore and PISO profile while a lower density is obtained for the 240 profile because of the different shape of this profile. The determination of the local halo density $\rho_{\odot,\text{Halo}}$ is performed under the requirement of the fulfillment of the local rotation velocity $v_\odot$. Therefore, a high maximal density for the inner ring leads to a low local DM halo density in order to keep the centripetal force at the Sun constant. This effect is reflected in the low total Galactic mass, which strongly depends on $\rho_{\odot,\text{halo}}$, and a high mass of the substructure which is about 30% of the total mass. The total mass distributions of the different halo profiles are shown in Figure 4.7b. There it becames apparent that the total mass distribution below a Galactocentric distance of one kpc is dominated by the visible matter due to the high density in the Galactic bulge. Two bumps are visible at about 5 and 15 kpc which are the influence of the DM rings. The behaviour of the integrated halo density becomes visible as soon as the influence of the outer ring disappears. The slope of the density decrease at large radii is the same for all considered cuspy profiles ($\propto 1/r^3$) which is the reason why their mass distributions show the same increase for radii above 30 kpc. However, the mass distribution of the PISO profile shows a steeper linear increase which result from its lower density decrease at large radii ($\propto 1/r^2$). The density distribution of the 240 profile show a rather steep decrease at large radii ($\propto 1/r^4$) than the cuspy profiles. For this reason its mass distribution decreases very fast and the total mass obtained from this profile is nearly constant for large radii. The velocity distributions in the Galactic disc and in the Galactic halo are presented in the Figures 4.7c and 4.7d. The strong influence of the halo substructure in the SP model is once more reflected in the RC within the Galactic disc. All profiles agree with the measurements of the rotation velocity within the Solar circle while in the radial region between the rings the rotation velocity is much decreased by the dense outer ring. The rotation curves of halo stars are obtained to be flat, structureless and in agreement with the experimental determination as shown in Figure 4.7d. The vertical gravitational potential at the position of the Sun is shown in Figure 4.7e. Its slope is closely connected to the local surface density as presented in Section 3.1.4. However, the local surface densities of all considered profile, as presented in Table 4.3, are too high in order to be consistent with the experimentally determined value of 71 ± 6 $M_\odot$ pc^{-2} which is also reflected in the steep increase of the vertical gravitational potential at the position of the Sun. The HWHM of the Galactic gas distribution was calculated for all density distributions according to Eq. (3.32). The results are presented in Figure 4.7f. The HWHM obtained from Moore and the NFW profile are in good agreement with the experimental data while the remaining profiles yield a strong decrease of the HWHM. For all density distributions a velocity dispersion of about 12 km s^{-1} is obtained which is inconsistent with the experimental determination of 8 ± 1 and 7 ± 1 km s^{-1} (see Section 3.1.4).

Summarizing the results in this section, the SP density model corresponds to our previous analysis of the diffuse Galactic gamma radiation [16]. There, the Galactic DM component

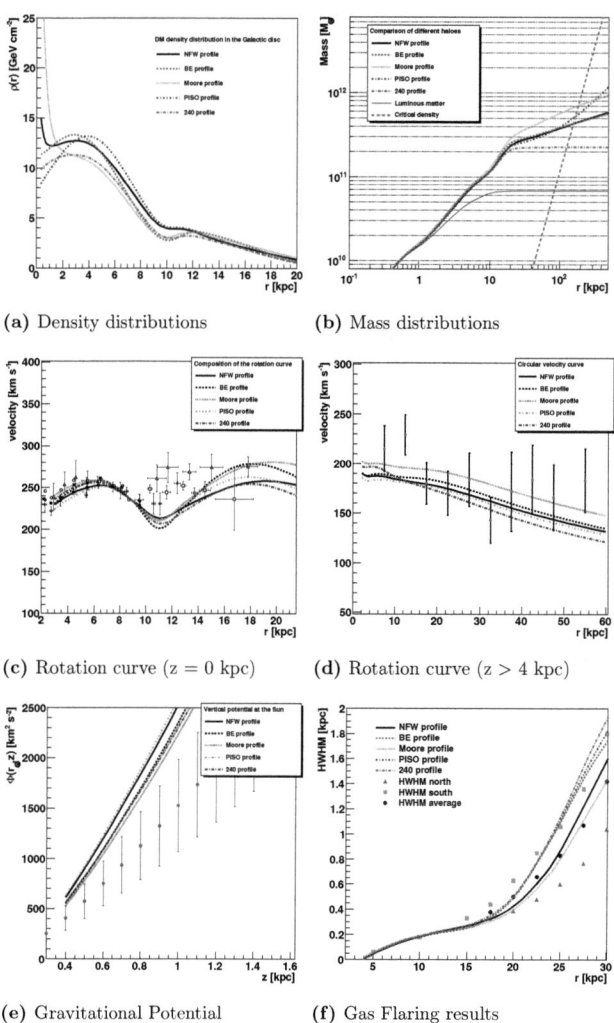

Figure 4.7: Resulting distributions of the SP density model. In this model the diffuse and the clumpy DM component are distributed according to the same density distribution. The radial dependence of the DM in the Galactic disc is shown in (**a**) while the mass distribution is given in (**b**). The velocity distribution in the Galactic disc and in the Galactic halo are shown in (**c**) and (**d**). In (**e**) the gravitational potential perpendicular to the Galactic plane is shown and in (**f**) the half-width-half-maximum of the Galactic gas distribution is given.

Astronomical constraint	M_{60kpc}	M_{tot}	$\rho_{\odot,tot}$	$\Sigma_{\odot,tot}$
Units	$10^{11}\ M_\odot$	$10^{12}\ M_\odot$	$M_\odot\ cm^{-3}$	$M_\odot\ cm^{-2}$
Experimantal values	4.0 ± 0.7	$1.0^{+0.3}_{-0.2}$	0.1 ± 0.01	71 ± 6
NFW	2.6	0.42	0.179	139.7
BE	2.7	0.43	0.150	135.1
Moore	4.0	0.48	0.165	133.85
PISO	3.0	0.50	0.189	142.85
240	1.6	0.22	0.170	127.65

Table 4.3: Astronomical properties of the considered halo profiles of the SP density model. For all considered halo profiles high total matter densities and high surface densities at the position of the Sun are obtained. The reason are the high ring densities obtained from the fit of the profiles to the longitudinal distribution of the gamma rays.

was assumed to be smoothly distributed in the Galactic halo which leads to a gamma ray signal from DMA which is proportional to the line-of-sight integral of the squared density distribution. In case of a clumpy DM component the gamma ray flux is dominated by the signal produced in the clumps. Then, the observed gamma ray flux is proportional to the line-of-sight integral of the linear density distribution. Hence, the densities of the DM rings have to be much increased compared to the results in [16] in order to find a good description of the spatial distribution of the gamma ray fluxes which is inconsistent with the astronomical observations. This effect can be vanished in a density model in which the density distributions of the two DM components are treated differently. Such a model is discussed in the subsequent section.

Double Profile density model

The analysis of the directionality of the diffuse Galactic gamma rays is performed in the way as for the SP model, i.e. that all halo slopes and eccentricities were fixed to the values obtained in [16], the Galactocentric distance to the Sun is $r_\odot = 8.3$ kpc and the survival probability of clumps is assumed to be constant. However, contrary to the SP model in this case the χ^2 function in Eq. (4.3) is minimised for different combinations of the halo profiles defined in Section 3.2.2.
The normalisation of the diffuse halo profile $\rho_{\odot,Halo,diff}$ is calculated under the requirement of the fulfillment of a total Galactic mass of $1 \cdot 10^{12}$ solar masses at a Galactocentric distance of 200 kpc. If the diffuse component is distributed according to the Moore profile the rotation velocity of the Sun is already given by the diffuse component and a clumpy DM component cannot be included anymore. In case of the 240 profile for the diffuse DM component a large for $\rho_{\odot,Halo,diff} = 2.2$ GeV cm^{-3} is obtained. The reason for this high normalisation is the strong decrease ($\propto 1/r^4$) of the density at large radii and the fixed value of its integral

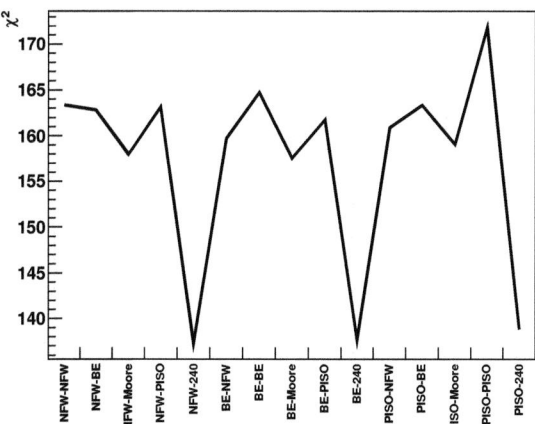

Figure 4.8: Fit results of the spatial dependence of the gamma ray flux from DMA for the fifteen considered profile combinations. The convention for the labels of the x-axis is that the first profile corresponds to the diffuse DM component and the second one is the halo profile of the clumpy component.

up to a radius of 200 kpc. As the required mass for the diffuse DM component and the local rotation velocity constraint cannot be simultaneously fulfilled by these profiles the halo combinations with a Moore or a 240 halo profile for the diffuse DM component are left out of the determination of the DM density distribution. For the remaining fifteen profile combinations the spatial dependence of the annihilation flux are analysed. Here, the survival probability of the DMCs in the MW, as described in Chapter 2, is set to be constant in the whole Galaxy, i.e. no tidal disruption of DMCs. This effect will be subsequently explained. The results of the approximation of the spatial distribution of the DMA flux are shown in Figure 4.8. The gamma radiation flux from the diffuse DM component results to be small compared to the gamma radiation flux from the clumpy component. This effect stems from the different scaling factor for these components. The scaling factor for the gamma ray flux of the diffuse component is per definition equal to unity whereas the scaling factor of the clumpy component (boost factor) is left free for the minimisation because of the unknown luminosity of DMCs. For that reason the determination method is sensitive to the gamma ray flux from the clumpy DM component. For all density profile combinations it therefore holds a simple principle: Similar halo profiles for the clumpy DM component lead to similar ring profiles. This principle is reflected in Figure 4.8 where it is shown that the χ^2 values of the approximation of the directionality of the DMA flux are more or less independent of the diffuse halo profile. Furthermore, it is clearly visible that combinations with the 240 profile for the clumpy DM component yield the best results for the approximation of the directionality of the diffuse Galactic gamma radiation. For this reason the analysis is restricted to profile combinations with the 240 profile for the clumpy component. The parameter settings of these combinations are shown in Table 4.4 while the parameters of the remaining combinations are given in Table D.1 in Appendix D. The longitudinal distribution of the gamma ray fluxes above a photon energy of 500 MeV obtained from the NFW-240 profile combination is shown in Figure 4.9 while the results of the other profile combinations are also presented in Appendix D. The total DM density distributions in the Galactic disc of the considered halo profile combinations are shown in Figure 4.10a. The shape of the DM rings are very similar for all three profiles. There the principle mentioned above is reflected again. The distributions of the cuspy profiles are equivalent except for the innermost region where the NFW profile of the diffuse component shows a much stronger increase than the BE profile. The DM density of the clumpy halo profile and the ring masses are about three times higher for the PISO-240 than for the other combinations. This difference as well as the equality of the density distribution for the cuspy profiles results from the different density decrease of the diffuse halo profiles. As the density of the PISO profile decreases proportional to $1/r^2$ at large radii its mass increases stronger than for the cuspy profiles which are proportional to $1/r^3$ at large Galactocentric distances. Consequently, the local density of the PISO profile is lower than for the other ones. This effect leads to a higher local density of the halo profile of the clumpy component and the DM rings and a lower boost factor, as shown in the total density at the Sun in Table 4.5. However, the fit to the longitudinal distribution is not effected. The difference of the NFW and the BE profile are hardly reflected in the resulting parameters since the total density in the GC is dominated by the baryonic matter component. The total Galactic mass is shown in Figure 4.10b. It shows a similar distribution for all halo profile combination which is an effect of the method used for the analysis. The obtained values for the total mass within 60 kpc and the total Galactic mass are given in Table 4.5. Both constraints are fulfilled by the considered halo combinations. The similarity of the NFW-240 and the BE-240 combination as well as the high density obtained for the clumpy component of the PISO-240 distribution is also apparent in the velocity distributions in the Galactic disc and in the halo, presented in the Figures 4.10c,d. The shape of the RC in the inner part of the Galctic disc is dominated by the disc component. Here, all combinations agree with the experimental values. At the outer Galaxy the NFW-240 and BE-240 profile yield too low velocities while a velocity increase results for the PISO-240 combination due to the higher mass of the outer ring. The circular rotation curve of halo stars are obtained to be structureless and flat in all cases. The slope of the vertical gravitational potential at the Sun's position is proportional to the local surface density. The results of the cuspy profiles show similar results with a less steeper slope than the experimental data. The resulting surface densities are compatible with the experimental determinations in Table 4.5. The PISO-240 profile provides a stronger potential with a steeper increase. For this reason a

	NFW - 240	BE - 240	PISO - 240
$\rho_{\odot,\text{diff}}$ [GeV cm^{-3}]	0.304	0.348	0.167
r_0 [kpc]	8.3	8.3	8.3
α_{diff}	1.0	1.0	2.0
β_{diff}	3.0	3.0	2.0
γ_{diff}	1.0	0.3	0.0
a_{diff} [kpc]	20.0	10.2	5.0
$\epsilon_{xy,\text{diff}}$	1.0	1.0	1.0
$\epsilon_{z,\text{diff}}$	1.0	1.0	1.0
$\rho_{\odot,\text{clump}}$ [GeV cm^{-3}]	0.0249	0.0245	0.0735
α_{clump}	2.0	2.0	2.0
β_{clump}	4.0	4.0	4.0
γ_{clump}	0.0	0.0	0.0
a_{clump} [kpc]	4.0	4.0	4.0
$\epsilon_{xy,\text{clump}}$	0.85	0.85	0.85
$\epsilon_{z,\text{clump}}$	0.7	0.7	0.7
ϕ_{gc} [°]	66	67	66
ρ_{IR} [GeV cm^{-3}]	2.6	2.56	7.9
R_{IR} [kpc]	3.5	3.4	3.4
$\sigma_{r,\text{IR}}$ [kpc]	3.5	3.6	3.6
$\sigma_{z,\text{IR}}$ [kpc]	0.312	0.302	0.315
$\epsilon_{xy,\text{IR}}$	0.85	0.85	0.85
ϕ_{IR} [°]	-121.5	-116	-123
M_{IR} [10^9 M$_\odot$]	7.73	7.3	23.2
ρ_{OR} [GeV cm^{-3}]	0.75	0.78	2.15
R_{OR} [kpc]	12.7	12.7	12.7
$\sigma_{r,\text{OR}}$ [kpc]	3.85	2.9	3.76
d_{OR}	4.0	4.0	4.0
$\sigma_{z,\text{OR}}$ [kpc]	0.8	0.83	0.8
$\epsilon_{xy,\text{OR}}$	0.95	0.95	0.95
ϕ_{OR} [°]	-113	-76	70
M_{OR} [10^{10} M$_\odot$]	2.0	1.6	5.47
boost factor	258	254	89
χ^2_{long} / d.o.f.	138.5 / 157	137.4 / 157	138.4 / 157
probability [%]	84.0	85.5	84.0

Table 4.4: Parameter settings of the DP model for the best approximations of the various DM halo profiles to the measured diffuse Galactic gamma radiation from EGRET.

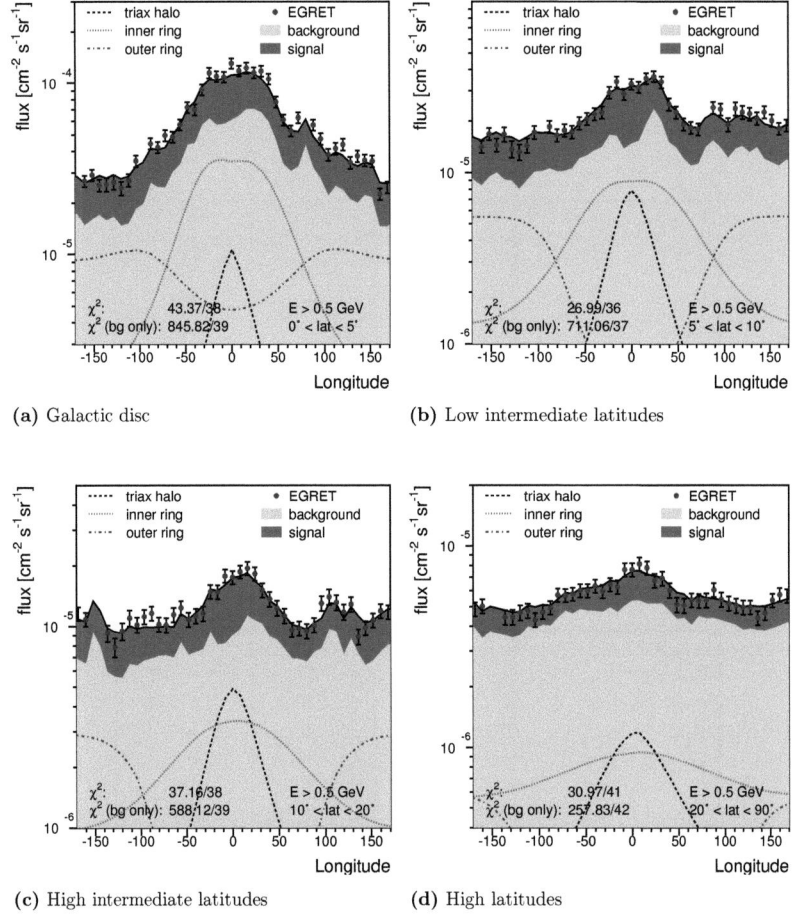

Figure 4.9: Longitudinal distribution of the diffuse Galactic gamma radiation obtained from the NFW-240 profile combination with a large scale structure of two rings at 3.5 and 12.7 kpc.

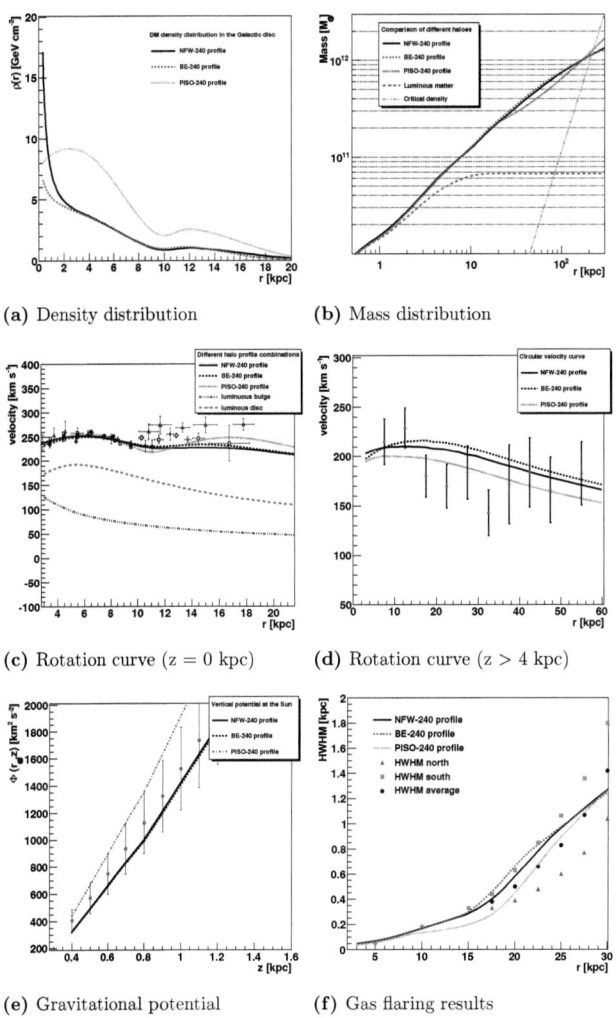

(a) Density distribution
(b) Mass distribution
(c) Rotation curve (z = 0 kpc)
(d) Rotation curve (z > 4 kpc)
(e) Gravitational potential
(f) Gas flaring results

Figure 4.10: Resulting distributions of the SP density model. In this model the diffuse and the clumpy DM component are distributed according to the same density distribution. The radial dependence of the DM in the Galactic disc is shown in (a) while the mass distribution is given in (b). The velocity distribution in the Galactic disc and in the Galactic halo are shown in (c) and (d). In (e) the gravitational potential perpendicular to the Galactic plane is shown and in (f) the half-width-half-maximum of the Galactic gas distribution is given.

Astronomical constraint	M_{60kpc}	M_{tot}	$\rho_{\odot,tot}$	$\Sigma_{\odot,tot}$	M_{clump}/M_{diff}
Units	10^{11} M$_\odot$	10^{12} M$_\odot$	M$_\odot$ cm^{-3}	M$_\odot$ cm^{-2}	%
Experimantal values	4.0 ± 0.7	$1.0^{+0.3}_{-0.2}$	0.1 ± 0.01	71 ± 6	-
NFW-240	4.6	1.06	0.106	80.2	3.3
BE-240	4.8	1.07	0.107	80.5	2.9
PISO-240	3.7	1.2	0.157	107.5	8.9

Table 4.5: Astronomical properties of the considered halo profiles of the DP density model. The total Galactic mass, the total matter density and the surface density near the Sun obtained from the NFW-240 and BE-240 profile combinations are in agreement with the experimental observation while the PISO-240 combination is in conflict with the local density. The ratio of the total mass of the diffuse DM component and the total mass of the clumps shows that the mass of the clumps in the PISO-240 profile is relatively increased by a factor of three compared to the other profiles. Nevertheless, in all cases the total mass is dominated by the diffuse DM component.

higher surface density is obtained which is not compatible with the data. The stronger vertical gravitational potential is also obtained in the estimation of the HWHM of the Galactic gas distribution shown in Figure 4.10f. For all halo combinations a velocity dispersion of $\sigma = 7.8$ km s^{-1} is obtained which is in good agreement with the experimental measurements in Section 3.1.4.

Summarizing the abovementioned results it becomes apparent that the best fit of the longitudinal distribution of the diffuse Galactic gamma radiation measured with EGRET is obtained for a 240 profile in combination with two DM rings at 3.5 and 12.5 kpc. Whether the astronomical constraints can be fulfilled or not depends on the density distribution of the diffuse DM component. A strong density decrease ($\propto 1/r^4$) for large Galactocentric distances is incompatible with the total Galactic mass while a low density increase ($\propto 1/r^2$) for large radii yields to high ring densities which are incompatible with the total matter density and the surface density at the position of the Sun. However, the resulting profile combinations show a low velocity distribution at the outer Galaxy which probably results from the assignment of the DM rings to the clumpy DM component. A diffuse fraction of the DM rings is discussed in the following section.

4.2.3 Clumpiness of the Dark Matter rings

It was shown above that an NFW-240 profile combination with two DM rings at about 3.5 and 12.5 kpc yield a good description of the longitudinal distribution of the diffuse gamma radiation fluxes. Furthermore, this model is in agreement with measurements of the local surface density, the total Galactic mass at a Galactocentric radius of 60 kpc, the circular

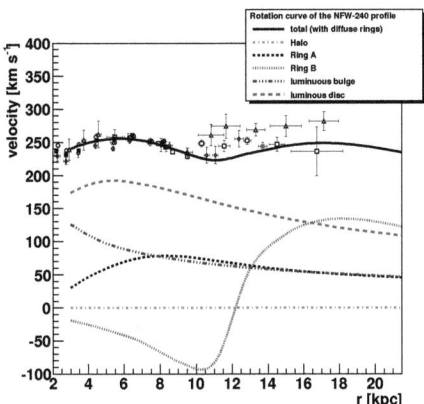

Figure 4.11: Galactic rotation curve with a fraction of 30% of diffuse DM in the outer ring.

rotation curve of stars in the Galactic halo and the HWHM of the gas distribution of the Galactic disc. However, the rotation velocity at the outer Galaxy is obtained to be too low as shown in Figure 4.10c. In the DP density model the DM rings contribute to the clumpy DM density distribution since they are assumed to consist of corotating DMCs and tidal streams which were likely produced by the infall of a dwarf galaxy. The smooth DM contribution of the rings is assumed to be diffused out of the ring. The inconsistency with the velocity distribution at the outer Galaxy indicates that a smooth component of the outer rings exist, which do hardly contribute to the gamma ray signal from DMA. In order to find the fraction of diffuse DM in the rings the DM density distribution is modified to

$$\rho_{\chi,\text{tot}} = \rho_\chi^{\text{Halo,diff}} + \eta_{IR} \cdot \rho_\chi^{\text{diff,IR}} + \eta_{OR} \cdot \rho_\chi^{\text{diff,OR}} + \rho_\chi^{\text{Halo,clump}} + (1.0 - \eta_{IR}) \cdot \rho_\chi^{\text{clump,IR}} + (1.0 - \eta_{OR}) \cdot \rho_\chi^{\text{OR}}. \quad (4.7)$$

In Section 3.3.2 a good description of the RC is obtained for the ring parameters given in Table 3.3.2. The maximal ring densities ρ_{IR} and ρ_{OR} found for the NFW-240 profile combination were increased to the values given in Table 3.3.2 and the fit to the EGRET data was performed again. The resulting velocity distribution is shown in Figure 4.11. A good fit is obtained for $\eta_{IR} = 0.0$ and $\eta_{OR} = 0.3$. Consequently, the RC prefers that the outer ring consists of 30% diffuse DM while the inner ring entirely consist of DMCs. The reason for this difference may be explained by the different age of the rings. If the infall of the dwarf galaxy occured late in the history of the MW the diffuse fraction brought in by the dwarf galaxy might not be completely diffused out of the ring.

P(0) [%]	75.0	50.0	25.0	0.0
only boost factor fitted				
χ^2_{long} / d.o.f.	142.2 / 157	145.7 / 157	150.12 / 157	155.3 / 157
probability [%]	77.9	71.1	61.8	50.0
fitted parameters				
ρ_{IR} [GeV cm^{-3}]	2.6	2.7	2.65	2.7
R_{IR} [kpc]	3.4	3.4	3.3	3.3
$\sigma_{\text{r,IR}}$ [kpc]	3.5	3.5	3.5	3.5
$\sigma_{\text{z,IR}}$ [kpc]	0.309	0.297	0.310	0.303
M_{IR} [10^9 M$_\odot$]	7.42	7.48	7.53	7.35
ρ_{OR} [GeV cm^{-3}]	0.74	0.74	0.735	0.715
R_{OR} [kpc]	12.7	12.7	12.8	12.8
$\sigma_{\text{r,OR}}$ [kpc]	3.7	3.9	3.7	3.7
$\sigma_{\text{z,OR}}$ [kpc]	0.8	0.8	0.8	0.8
M_{OR} [10^{10} M$_\odot$]	1.9	1.9	1.8	1.8
χ^2_{long} / d.o.f.	139.9 / 157	141.3 / 157	146.1 / 157	148.24 / 157
probability [%]	81.6	79.5	61.8	65.9

Table 4.6: Parameter settings of the DP model for the best approximations of the various DM halo profiles to the measured diffuse Galactic gamma radiation.

Survival probability of Dark Matter clumps

The analysis of the diffuse Galactic gamma radiation for the SP and the DP density model is performed with a constant survival probability for DMCs which means that the cores of DM subhaloes are not destroyed by the tidal forces. In this section the influence of the central value of the survival probability on the resulting density distributions is examined. The fit of the NFW-240 profile combination to the longitudinal distribution of the diffuse Galactic gamma rays is performed again for the $P(0) = 0.75, 0.5, 0.75$ and 0.0. First all parameters except for the boost factor were fixed in order to obtain comparable results to $P(0) = 1.0$. In the second step the ring parameters were left free for the fit to the longitudinal distribution of the gamma rays. The results are summarised in Table 4.6. In case of a fixed profile it is shown that the fit of the longitudinal distribution of the gamma radiation mildly depends on the survival probability of clumps. A decline of the χ^2 value is obtained but in all cases the results are compatible with the experimental data. This effect results from the radial dependence of the survival probability shown in Figure 4.3. It influences only the inner part of the Galaxy while the gamma ray fluxes at large Galactic longitudes and at the Galactic poles are not affected. If the parameters of the DM rings are left free the fit can be improved by the increase of the inner ring which compensates the decrease of the central value of the survival probability. The mass of the outer ring stays unchanged because at radii larger than 5 kpc the survival probability is constant. In case of $P(0) = 0$ the compensation of the survival probability by the inner ring seems not effective anymore. In this case the ring mass is reduced in order to increase the gamma radiation flux from the halo of the clumpy DM component. Nevertheless, a mild dependence of the fit results concerning the variation of the central value of the survival probability of DMCs is found which results from the special radial dependence of the survival probability.

4.3 Discussion

In the last sections the results of the analysis of the diffuse Galactic gamma radiation were presented. It was shown that the SP density model, in which a single density distribution is assumed for the diffuse and clumpy DM component, is not able to describe the astronomical constraints because of the high local DM density originating from the high ring masses obtained in the this model. The reason for the high ring masses is the fit to the gamma radiation at large longitudes and low latitudes and the domination of the gamma ray flux from DMA in the clumps.

Different density distributions for the diffuse and the clumpy DM distribution were used in the DP model. In this model the total Galactic mass is mainly provided by the diffuse DM component which, however, hardly contributes to the diffuse Galactic gamma radiation. The gamma radiation from DMA is mainly produced by the clumpy DM component. Thus, in the DP model the total Galactic mass is decoupled from the DMA signal. In this case lighter rings are obtained which are consistent with the local surface density and the local matter density. For a diffuse DM fraction of about 30% (0%) in the outer (inner) ring is compatible with the velocity distribution at the outer disc. This difference between inner and outer ring may be due to the different ages, resulting in a larger diffusion at the diffuse DM component. For the whole Galaxy the diffuse component is more than 90% of its mass (see Table 4.5). The rings contribute 0.7% (1.9%) of the total mass of the inner (outer) ring, respectively.

Towards the GC the DMCs may have experienced more tidal disruption by the collision with stars or simply by passing the Galactic disc. However, the mass of the inner ring is strongly correlated with the survival probability of the clumps, so the survival probability is not well determined.

For the boost factors needed for the diffuse Galactic gamma rays Galactic propagation models can be used to estimate the local Galactic antiproton flux. Simple propagation models overestimate the Galactic antiproton flux by an order of magnitude as discussed in [17]. However, recent propagation models are afflicted with large uncertainties because of the little-known parameters of the local variation of the Galactic magnetic field or trapping processes in magnetic clouds. In [151] it was shown that the overestimation of the antiproton flux can be reduced in a more detailed propagation model including Galactic winds.

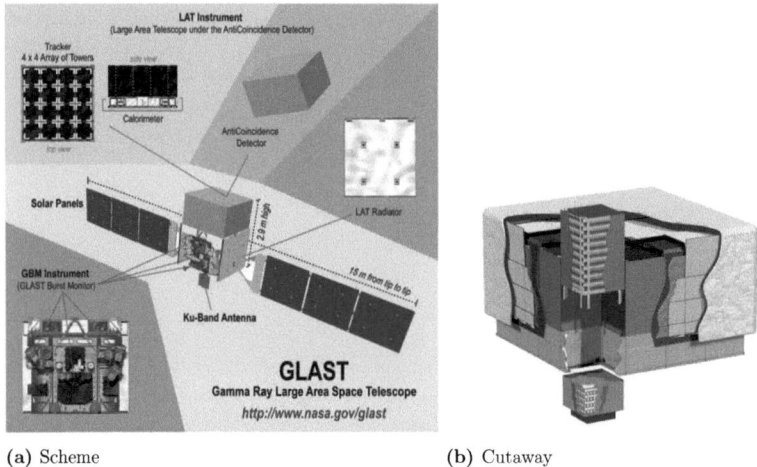

(a) Scheme (b) Cutaway

Figure 4.12: (a) A schematic picture of the Fermi satellite is shown. The different components of the Large Area Telescope are presented at the top of the picture while the instruments of the Gamma-Ray Burst Monitor are shown at the lower left edge. (b) A cut-away picture of the LAT instrument is shown. The LAT is enclosed by the segmented anticoincidence detector which is yellow coloured. One tower of the 4×4 array is depicted in more detail. The upper part consists of a silicon strip detector while the calorimeter is located at the lower part of the tower.

4.4 Recent and future gamma ray observations

On June 11th 2008 the Gamma-Ray Large Area Space Telescope (GLAST), was succesfully launched at the Cape Canaveral Air Force Station in Florida and later renamed the Fermi Gamma-Ray Space Telescope. It is the successor of the EGRET telescope. The main aims of the Fermi Telescope are the identification of so far unidentified point sources, the discovery of new point sources and the observation of so-called gamma-ray bursts. Two instruments are installed at the Fermi satellite - the Large Area Telescope (LAT), which is the main instrument, and the Gamma-Ray Burst Monitor (GBM). In Figure 4.12a the design of the Fermi satellite is shown. The LAT instrument is located on top of the satellite while the GBM is a system of twelve small NaI detectors (bunches of three detectors on each corner) at the lower part of the satellite. Large area observations of the Galactic gamma radiation are done with LAT. It is sensitive to gamma rays in the energy range of 20 MeV to 300 GeV and has an on-axis effective area of about 8000 cm^2 for $E > 1$ GeV. The measurements of the GBM complement the LAT in its observation of point sources and is sensitive to x-ray and

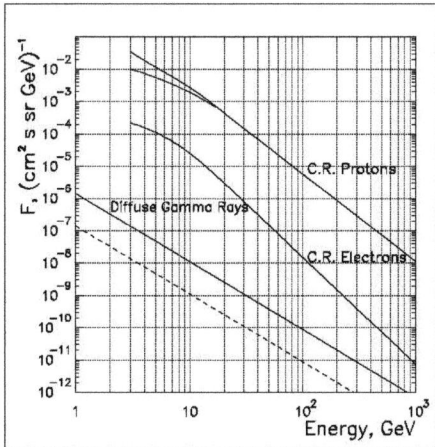

Figure 4.13: Differential spectra of primary cosmic rays and the diffuse Galactic gamma radiation are shown. Below an energy of about 15 GeV the spectrum of the CR protons is affected by the protons emitted by the Sun. The proton spectrum for the maximal and minimal Sun activity is shown. The fluxes of the extragalactic diffuse gamma radiation are about five orders of magnitude smaller than the proton fluxes. The dashed line represents 10% of the extragalactic background radiation spectrum. Figure taken from [153].

gamma rays in the energy range of 8 keV to 40 MeV. The measurement of the diffuse Galactic gamma radiation is performed with the LAT instrument which is shown in more detail in Figure 4.12b. It is designed as a pair-production telescope composed of a 4×4 grid of towers. Each tower consists of a silicon strip detector and a tungsten-foil tracker/converter, mated with a hodoscopic cesium-hiodide calorimeter. The grid of towers is covered by a segmented plastic scintillator anticoincidence detector in order to reject charged-particle backgrounds from CR. The maximal ratio of CR electron flux to the averaged photon flux in the energy range of 3 to 10 GeV is about 3000 while the CR proton flux is about four to five orders of magnitudes higher than the photon flux as shown in Figure 4.13. The anticoincidence detector of the Fermi/LAT instrument was designed for an efficiency of at least $3 \cdot 10^{-4}$ because of the low efficiency of the calorimeter to capture the CR proton energy. Electrons from CR produce an energy deposit in the calorimeter which is similar to the deposit of the gamma radiation. Consequently, the anticoincidence detector of the Fermi/LAT mainly suppresses the contamination with CR electrons while a combination of calorimeter and anticoincidence detector is used to reject CR protons [152]. The segmentation of this detector is designed to suppress self-vetoes produced by the backsplash effect in which secondary photons from the

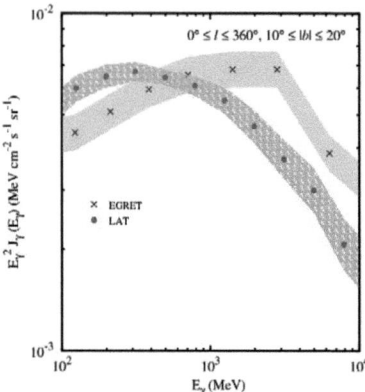

Figure 4.14: Preliminary energy spectrum of the diffuse Galactic gamma radiation averaged over all Galactic longitudes in the latitudinal region of $10° < |b| < 20°$. The EGRET data is shown in blue crosses and the Fermi/LAT data is shown in red dots. Systematic uncertainties are presented as shaded areas. Adapted from [155].

electromagnetic shower created by incident high-energetic photons can Compton scatter in the anti-coincidence detector and produce veto signals from the recoil electrons.

After launch Fermi was calibrated for two months. Then gamma radiation data were recorded one year. Recently, this data were made publicly available on the webpage of the Fermi Science Support Center (FSSC) [154]. It shows the diffuse Galactic gamma radiation in combination with gamma ray emission from Galactic point sources. In the intermediate latitude region (region D in Table 4.2.1) the number density of point sources is small and the diffuse Galactic gamma radiation is the dominant contribution. Therefore, the gamma radiation from this region was analysed in [155]. The resulting energy spectrum was produced using the instrument response function P6_V3_DIFFUSE and is shown in Figure 4.14. The EGRET excess above a photon energy of 1 GeV is not confirmed by this data. Analyses of the diffuse Galactic gamma radiation measured with Fermi/LAT are not published so far, but a preliminary model of the diffuse gamma radiation is publicly available [154]. There measurements of the gamma radiation of nearly 10 months were used and the exposure and point spread function maps were generated using the P6_V3_DIFFUSE instrument response function. The point sources were taken from 9 months data using a preliminary model of the diffuse emission. The all-sky map of the observed gamma ray counts is shown in Figure 4.15a. The diffuse Galactic gamma radiation was modelled using contributions from neutral pion decay, bremsstrahlung and inverse Compton scattering. For the calculation

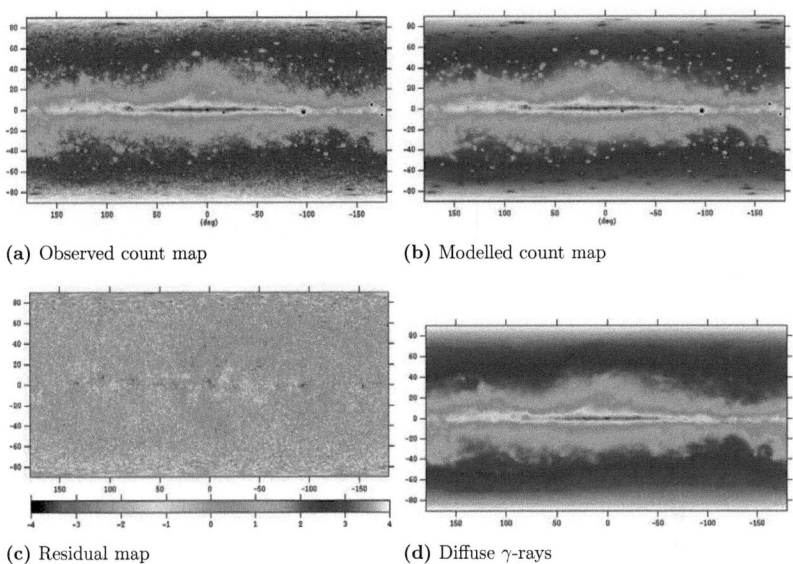

Figure 4.15: All skymaps are produced for gamma ray energies between 300 MeV and 20 GeV. The count map of the gamma ray measurements of nearly 10 months with Fermi/LAT is shown in **(a)** while the count map of the modelled diffuse gamma radiation in combination with the modelled point sources is given in **(b)**. In both pictures the counts are illustrated with the same logarithmic scale. In Figure **(c)** the residual map expressed in sigma values of $(N_{obs} - N_{pred})/\sqrt{N_{pred}}$ is given. It represents the statistical difference between (a) and (b). The diffuse model itself is shown in Figure **(d)**. It shows the integrated counts scaled by the exposure in the same scale as in (a) and (b). Figures taken from [156].

of these contribution measurements of the all-sky Leiden-Argentina-Bonn (LAB) composite survey [157] were used for the Galactic distribution of ionised hydrogen N(HI) and the velocity integrated intensities of carbon monoxide W(CO) were obtained from the Center for Astrophysics compilation [158]. Maps of these distributions were derived for 6 concentrical rings around the GC using the RC in [94]. The `GALProp` code was used to calculate the intensities of primary and secondary CR protons and electrons. Then all gamma radiation components were convolved with the LAT point spread function in the energy region from 0.3 to 20 GeV. Then the model prediction in combination with the modelled point sources, as shown in Figure 4.15b, was fitted to the observed data in the energy range between 120 MeV and 20 GeV. The resulting residual map and the model prediction of the diffuse Galactic gamma radiation are shown in Figures 4.15c,d. The resulting gamma radiation fluxes were stored in 30 logarithmically equidistant energy bins from 50 MeV to 100 GeV in the *gll_iem_v02.fit* file whereas the gamma ray fluxes above 20 GeV were extrapolated and globally normalised to fit the Fermi/LAT data. These fluxes do not contain the isotropic component produced by the EGBR. This contribution was separately determined and tabulated in the file *isotropic_iem_v02.txt*. This model is the most accurate description of the Fermi/LAT data which the LAT team produced so far but it shows some caveat like non-neglibible residuals between the model and the observed fluxes in the low latitude region, as shown in Figure 4.15c, and the extrapolation of the model above a photon energy of 20 GeV. A more detailed description of the model can be found at [156]. In the present thesis this model was used to analyse the Fermi/LAT data. It was announced by the FERMI collaboration that a new model for the diffuse Galactic gamma radiation will be published in January 2010 which is, however, too late for the consideration in this thesis.

The gamma ray fluxes from the *gll_iem_v02.fit* model were considered in the same way as the gamma radiation fluxes measured with EGRET. The considered data is restricted to the energy range from 120 MeV to 20 GeV since in this range the model was fitted to the data. A more detailed description can be found in the documentation to this model on the FSSC webpage.

It was shown in the previous section that an NFW profile for the diffuse DM component and a 240 profile in combination with two DM rings for the clumpy DM component yield a good approximation of the gamma radiation measured with EGRET. This profile combination is used for a fit to the preliminary Fermi/LAT data. The uncertainty of the shape of the energy spectrum is assumed to be 15% because of the possible background of CR protons. The resulting energy spectra in the considered sky regions, defined in Table 4.2.1, are presented in Figure 4.16 [1]. The extragalactic background radiation is parametrised with a simple power law. The parameters were adapted from [151]. The energy spectra in all regions show a significantly improved fit if a DM contribution with a WIMP mass of about 60 GeV is added. The DM contribution is decreased compared to the energy spectrum obtained from the EGRET data. This decrease is reflected in the boost factor which is reduced

---
[1]The fits to the Fermi/LAT data were performed in collaboration with Iris Gebauer.

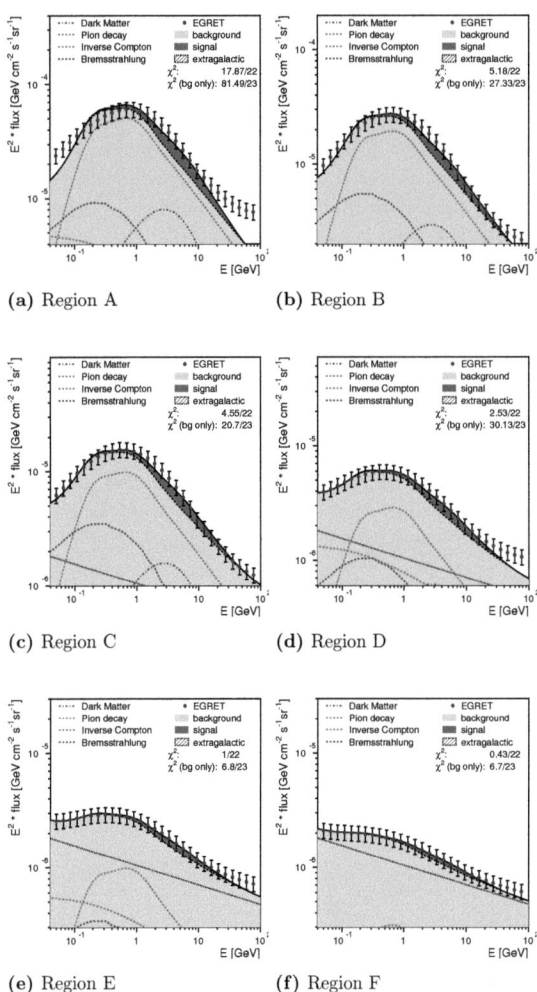

Figure 4.16: Preliminary energy spectra of the diffuse Galactic gamma radiation measured with Fermi/Lat in the regions defined in Table 4.2.1. The presented data was obtained from the *gll_iem_v02* model of the Galactic gamma emission. The uncertainty of the data is assumed to be 15%. The fitted line represents the same fit as to the EGRET data with a constant boost factor in all regions. The boost factor is reduced by a factor three compared to the fit to the EGRET data shown in Figure 4.4.

by a factor of about 3. The data points at high energies are understood as an additional contribution from CR protons [159]. At a photon energy of about 1 (10) GeV about 60% (0%) of the gamma rays split into an electron-positron pair in the calorimeter. Therefore, for larger energies the contribution of misinterpreted photons produced by CR protons is larger as shown by the tail in the CR proton flux shown in Figure 4.13. In this thesis the diffuse gamma radiation in the energy region above 20 GeV is not taken into account.

Next the longitudinal distribution of the gamma radiation measured with Fermi/LAT is considered. Here, the uncertainty of the shape is assumed to be smaller (7%) than for the energy spectrum since in this case the integrated flux in the longitudinal bins is considered, so the possible background at higher energies has a smaller weight. In order to check whether a ringlike substructure of the DM halo is needed to describe the gamma radiation with Fermi/LAT four different cases examined. The NFW-240 profile is considered with two rings, without the inner ring, without the outer ring and without both rings. The results are presented in Table 4.7. There it is shown that the best fit of the longitudinal distribution of the gamma ray fluxes is obtained for a halo profile combined with a substructure of two DM rings. Therefore, such a profile can be regarded as a generic DM profile for the DM density distribution by virtue of its consistency with recent astronomical observations and the measurements of the diffuse Galactic gamma radiation with Fermi/LAT. The longitudinal distribution of the gamma radiation fluxes obtained from the NFW-240 profile with two DM rings is shown in Figure 4.17.

4.4.1 Future indirect DM searches

The Alpha Magnetic Spectrometer (AMS) is a large acceptance, high precision superconducting magnetic spectrometer which is constructed, tested and operated by an international team composed of 56 institutes from 16 countries (including KIT). It is designed for an operation time of at least three years in a height of 400 km at the International Space Station (ISS). Its installation is planned for the last NASA space shuttle mission in the autumn of

Distribution	χ^2 / d.o.f.	boost factor
With rings	30 / 157	80
Without OR	80 / 157	100
Without OR	65 / 157	138
Without rings	175 / 157	264
Without DM	250 / 157	-

Table 4.7: Different combinations of the DM rings for the NFW-240 profile combination are shown. An additional DM contribution improves the fit while the best fit is obtained for two DM rings.

4.4 Recent and future gamma ray observations

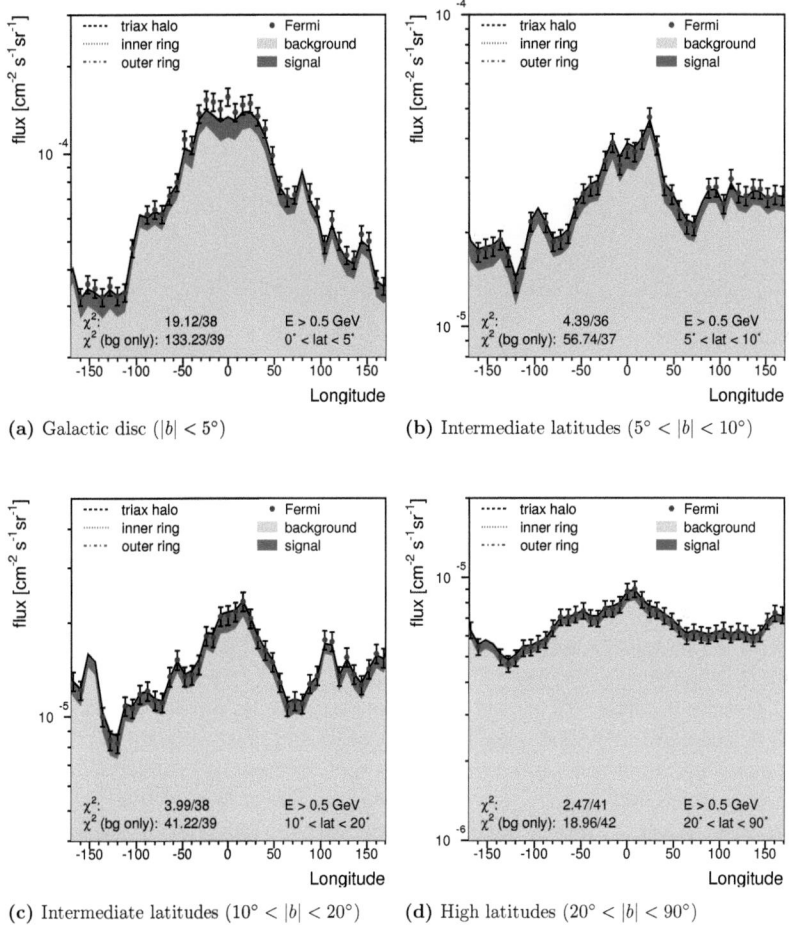

Figure 4.17: Preliminary longitudinal distribution of the diffuse Galactic gamma radiation measured with Fermi/LAT. In **(a)** the longitudinal distribution of the gamma radiation fluxes in the Galactic disc are presented. The background radiation is shown in yellow while the gamma ray flux from DM is presented in red. The longitudinal distributions in the intermediate latitude regions are presented in **(b)** and **(c)**. The gamma ray fluxes at large latitudes, where the influence of the extragalactic background is highest, is presented in **(d)**.

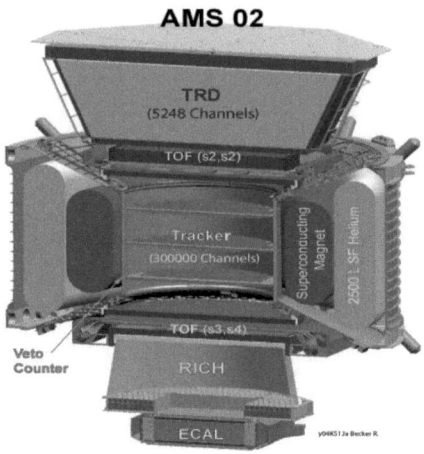

Figure 4.18: Schematic picture of the configuration of the AMS-02 detector.

2010. The main aim of the AMS project is search for antimatter and DM as well as the measurement of cosmic radiation spectra of elements up to $Z \lesssim 25$. Gamma rays can be measured up to a few hundred GeV. A schematic picture of the AMS-02 detector is shown in Figure 4.18. The main building blocks of the AMS detector are the Transition Radiation Detector (TRD), the Silicon Tracker (ST), the Time of Flight (TOF) spectrometer, the Ring Image Cherenkov Counter (RICH) and the Electromagnetic Calorimeter (ECAL).

The TRD is designed to separate signals from electrons and protons in order to distinguish signals from their antiparticles (positrons and antiprotons) from background radiation with a rejection factor of $10^{-3} - 10^{-2}$. The TOF system consists of four scintillator planes and provides a fast trigger to experiment concerning the up/down separation of traversing particles. In between the TOF system the ST is located. It is surrounded by a superconducting magnet which is cooled down by liquide helium to a temperature of 1.8 K and produces a magnetic field of about 0.87 T. This field changes the trajectory of traversing charged particles in order to precisely measure their rigidity and the sign of their electrical charge with the ST which consists of eight thin layers of double-sided silicon microstrip detectors. At the bottom of AMS the RICH detector and the ECAL are located. The RICH detector consists of a ~ 1.5 cm thick plane of radiator material (NaF) and a subjacent matrix of photomultipliers. In order to increase the efficiency of the RICH counter it is encircled by a mirror which is formed as a cone. In the middle of the photomultiplier matrix a hole to the ECAL, which is placed just below the RICH counter, is left open. The ECAL is used to discriminate electromagnetic and hadronic showers and to image their development in

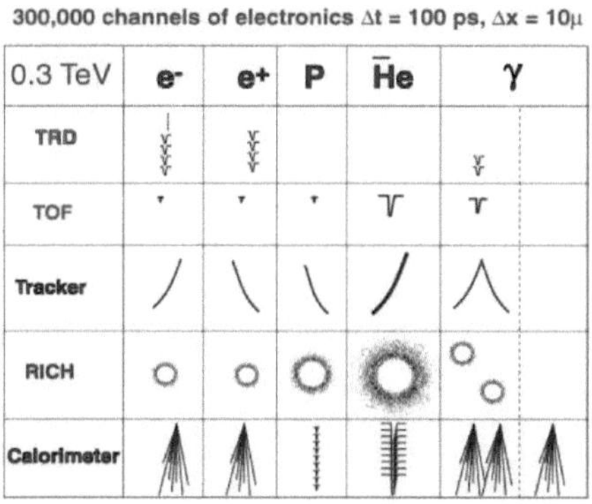

4.4 Recent and future gamma ray observations 109

Figure 4.19: Schematic picture of the particle identification signals in the AMS-02 detector.

3D. In addiation to all explained components an anticoincidence veto counter is installed between the magnet and the ST. It assures that only particles which pass through the entire detector are triggered [160, 161]. The signals of electrons, positrons, protons, anti-helium nuclii and photons in the different detector components are shown in Figure 4.19. The edge of AMS over EGRET and Fermi/LAT is the identification of particles in the TRD, ST, TOF and the RICH counter which further reduces the fraction of misinterpreted photon signal above an energy of 1 GeV produced by protons from cosmic radiation. The AMS mission will therefore play an important role in the indirect search for DM.

4
Summary and Outlook

In the first part of this thesis recent astronomical data on the rotation curve (RC), the local surface density and the total matter density near the Sun, as obtained from the movement of nearby stars, are used to constrain the local DM density. No sensitivity of these constraints to cuspy or cored profiles is found since the gravitational potential near the Galactic centre (GC) is dominated by the visible matter of the MW. The local DM density is found to be in the range of 0.2 to 0.4 GeV cm^{-3} for spherical haloes and a DM density up to 0.7 GeV cm^{-3} is obtained for oblate haloes including dark discs. However, the RC in the Galactic disc shows a change of slope at about 9 kpc, which cannot be described by a smooth DM halo profile. A DM halo profile in combination with a DM large scale structure composed of two doughnutlike DM rings at 4.5 kpc and 12.5 kpc yield a good description of the RC and is consistent with the other astronomical constraints, especially the flattening of the gas flaring between 10 and 20 kpc can only be explained by such a ringlike structure. Rings of DM are expected from the infall of a dwarf galaxy in the gravitational potential of the MW, as is known from numerical simulations [144].

In the second part of the thesis the diffuse Galactic gamma radiation measured with the EGRET satellite is used to constrain the DM density profile. Following the analysis in [16] the energy spectrum and the spatial distribution of the gamma radiation fluxes are simultaneously considered. In contrast to the study in [16] in the present analysis the DM component was assumed to be a composition of a smoothly distributed component, which describes the distribution of individual DM particles, and a component consisting of small subhaloes called DM clumps. The two components were distributed either according to the same profile, called Single Profile (SP) density model, or to different profiles, called Double Profile (DP) density model.

1. The density distribution in the SP model corresponds to the density distribution in [16] in which the diffuse Galactic gamma radiation is analysed using a smoothly distributed

DM component. If DMA is dominated by the clumpy component the signal is only linearly proportional to the clump density, not the average DM density squared. In this case a higher averaged DM density is needed, also in the rings, which leads to values of the local surface density and gravitational potential incompatible with the data.

2. In the DP density model the spatial distribution of the diffuse Galactic gamma radiation measured with EGRET is best described by a halo combination of a cored profile for the DM clumps and a cuspy profile for the diffuse DM component in agreement with recent high-resolution N-body simulations. The data need the additional ringlike structures with parameters in agreement with the ringlike structures from the astronomical data discussed above. The diffuse DM component makes up more than 90% of the total mass while the rings provide 0.7% – 1.9% of the total mass. The ratio of visible mass to DM mass is of the order 1/20 which is in good agreement with N-body simulations.

3. Preliminary data of the diffuse Galactic gamma radiation measured with Fermi/LAT show less of an excess. The resulting density distribution improves the χ^2 value of the fit of the longitudinal distribution of the gamma ray fluxes from 250 / 157 (without DMA) to 30 / 157 (with DMA). The intensity of the gamma radiation from DMA is decreased by a factor of three, resulting in a factor of three lower boost factor, but for the high statistics of the Fermi data the inclusion of a gamma ray signal from DMA in a description of the data yields a significant improvement.

4. In the present thesis a simple model for the propagation of cosmic ray particles is used. For such a model the Galactic antiproton flux from the annihilation of a WIMP with a mass below 100 GeV is overestimated by a factor of about 10, as described in [17]. This factor is reduced in a more sophisticated propagation model including Galactic winds and for recent observations by Fermi/LAT an overestimation by a factor of three is obtained [151] and can be further reduced by a more detailed 3D-modeling of the spiral arms.

The preliminary Fermi/LAT energy spectrum of the gamma rays is increasing for photon energies above 20 GeV which probably is the result of a further background from misinterpreted cosmic-ray protons (fake photons). This background is currently under discussion and a new release of the data with harder cuts against the background radiation is expected soon. An analysis of these data remains to be done.

A further reduction of the background radiation component from fake photons is likely to be reached with the AMS experiment which will be launched in 2010. In contrast to the EGRET and the Fermi/LAT detector AMS has a magnetic spectrometer combined with an electromagnetic calorimeter, a transition radiation detector and a Cherenkov counter. The combination of these detectors will allow a clean identification of all charged cosmic rays and gamma rays, thus providing unique data for the indirect search for DM.

A
GALPROP datacard for the Galactic background radiation

```
Title                 = conventional model
n_spatial_dimensions  = 2
r_min                 = 0.         min r (in kpc)
r_max                 = 20.        max r
dr                    = 1.         delta r
z_min                 = -4.        min z
z_max                 = +4.        max z
dz                    = 0.2        delta z
p_Ekin_grid           = Ekin       p or Ekin alignment
Ekin_min              = 1.e0       min kinetic energy per nucleon (MeV)
Ekin_max              = 1.e9       max kinetic energy per nucleon
Ekin_factor           = 1.2        kinetic energy per nucleon factor
E_gamma_min           = 1.e0       min gamma-ray energy (MeV)
E_gamma_max           = 1.e8       max gamma-ray energy (MeV)
E_gamma_factor        = 1.3        gamma-ray energy factor

integration_mode      = 1
nu_synch_min          = 1.e6       min synchrotron frequency (Hz)
nu_synch_max          = 1.e10      max synchrotron frequency (Hz)
nu_synch_factor       = 1.5        synchrotron frequency factor
long_min              = 0.25       gamma-ray intensity skymap long minimum (deg)
long_max              = 359.75     gamma-ray intensity skymap long maximum (deg)
lat_min               = -89.75     gamma-ray intensity skymap lat minimum (deg)
lat_max               = +89.75     gamma-ray intensity skymap lat maximum (deg)
d_long                = 0.5        gamma-ray intensity skymap long binsize (deg)
d_lat                 = 0.5        gamma-ray intensity skymap lat binsize (deg)

D0_xx                 = 5.8e28     diff coeff at reference rigidity
```

```
D_rigid_br           = 4.0e3     ref rigidity for diff coeff in MV
D_g_1                = 0.33      diff coeff index below reference rigidity
D_g_2                = 0.33      diff coeff index above reference rigidity
diff_reacc           = 1         1=include diff reacceleration
v_Alfven             = 30.       Alfven speed in km s-1
damping_p0           = 1.e6      MV - some rigidity (where CR density is low)
damping_const_G      = 0.02      a const derived from fitting B/C
damping_max_path_L   = 3.e21     Lmax~1 kpc, max free path
convection           = 0         1=include convection
v0_conv              = 0.        v_conv=v0_conv+dvdz_conv*dz (km s-1)
dvdz_conv            = 10.       v_conv=v0_conv+dvdz_conv*dz (km s-1 kpc-1)

nuc_rigid_br         = 9.e3      ref rigidity for nucl injection index (MV)
nuc_g_1              = 1.98      nucl injection index below ref rigidity
nuc_g_2              = 2.42      nucl injection index above ref rigidity
inj_spectrum_type    = rigidity  rigidity or beta_rig or Etot
                                 nucleon injection spectrum type

electron_g_0         = 1.6       el injection index below ref rigidity0
electron_rigid_br0   = 4.e3      ref rigidity0 for el injection index (MV)
electron_g_1         = 2.54      el injection index above ref rigidity0
electron_rigid_br    = 1.e9      ref rigidity for el injection index (MV)
electron_g_2         = 5.        el injection index index above ref rigidity

He_H_ratio           = 0.11      He/H of ISM, by number
X_CO                 = 1.9e20    conversion factor from CO integrated
                                 temperature to H2 column
                                 density (xcomode=0) (mol cm-2 K-1 (km/s)-1)
xcomode              = 1         0:X_CO constant 1:X_CO increasing
# 0.15*X_CO ( 0.0 kpc  < r < 4.0 kpc)
# 0.3 *X_CO ( 4.0 kpc  < r < 6.0 kpc)
# 0.4 *X_CO ( 6.0 kpc  < r < 8.0 kpc)
# 0.65*X_CO ( 8.0 kpc  < r < 10.0 kpc)
# 2.5 *X_CO ( 10.0 kpc < r < 15.0 kpc)
# 5.0 *X_CO ( 15.0 kpc < r )

fragmentation        = 1         1=include fragmentation
momentum_losses      = 1         1=include momentum losses
radioactive_decay    = 1         1=include radioactive decay
K_capture            = 1         1=include K-capture

start_timestep       = 1.e7
end_timestep         = 1.e2
timestep_factor      = 0.25
timestep_repeat      = 20        number of repeats per timestep in timetep_mode=1
timestep_repeat2     = 0         number of timesteps in timestep_mode=2
timestep_print       = 10000     number of timesteps between printings
timestep_diagnostics = 10000     number of timesteps between diagnostics
```

```
control_diagnostics    = 0              control detail of diagnostics

network_iterations     = 2              number of iterations of entire network

prop_r                 = 1              1=propagate in r
prop_x                 = 1              1=propagate in z
prop_p                 = 1              1=propagate in momentum

use_symmetry           = 0              0=no symmetry, 1=optimized symmetry
vectorized             = 0              0=unvectorized code, 1=vectorized code

source_specification   = 0              1:r,z=0 2:z=0
source_model           = 1              0=zero 1=parameterized 2=Case&B 3=pulsars
                                        4= 5=S&Mattox 6=S&Mattox with cutoff 7=Lorimer

B_field_model          = 050100020 bbbrrrzzz bbb=10*B(0) rrr=10*rscale zzz=10*zscale

proton_norm_Ekin       = 1.e+5          p kinetic energy for norm (MeV)
proton_norm_flux       = 5.e-9          flux of p at norm energy (cm-2 sr-1 s-1 MeV-1)
electron_norm_Ekin     = 34.5e3         el kinetic energy for norm (MeV)
electron_norm_flux     = .4e-9          flux of el at norm energy (cm-2 sr-1 s-1 MeV-1)

max_Z                  = 28             maximum number of nucleus Z listed
use_Z_1                = 1
...
use_Z_28               = 1

iso_abundance_01_001   = xxx            relative abundaces of primary spectrums
...                                     tuned to get propagated fluxes correct
...                                     different for optimized and conventional model
...                                     no big influence on gammas
...                                     see astro-ph/0101068 for details

total_cross_section    = 2              total cross sec option: 0=L83 1=WA96 2=BP01
cross_section_option   = 012            100*i+j i=1: use Heinbach-Simon C,O->B
                                        j=kopt j=11=Webber, 21=ST
t_half_limit           = 1.e4           year - lower limit on radioactive half-life
                                        for explicit inclusion

primary_electrons      = 1
secondary_positrons    = 1
secondary_electrons    = 1
secondary_antiproton   = 2
tertiary_antiproton    = 1
secondary_protons      = 1
gamma_rays             = 1              1=compute gamma rays
IC_anisotropic         = 0              1=compute anisotropic IC
synchrotron            = 0              1=compute synchrotron
```

B
Rotation velocities

In this appendix the experimental data for the velocity distribution within the Galactic disc of the MW is presented. First the data sets adapted from different publication [93–96,98,99,162] are shown in Tables B.1 through B.7, then the averaged velocities are shown in the Tables B.8 to B.14. All data sets are scaled to a Galactocentric distance of the Sun $r_\odot = 8.3$ kpc and a local rotation velocity of $v_\odot = 244$ km s^{-1} and then averaged according to

$$r = \sum \frac{r_i}{\sigma_{r,i}^2} / \sum \frac{1}{\sigma_{r,i}^2}, \qquad \sigma_r = \frac{1}{\sqrt{\sum 1/\sigma_{r,i}^2}} \qquad (B.1)$$

$$v = \sum \frac{v_i}{\sigma_{v,i}^2} / \sum \frac{1}{\sigma_{r,i}^2}, \qquad \sigma_v = \frac{1}{\sqrt{\sum 1/\sigma_{v,i}^2}} \qquad (B.2)$$

r [kpc]	σ_r [kpc]	v [km s^{-1}]	σ_v [km s^{-1}]
2.92	0	239.12	16
3.75	0	253.76	15
4.58	0	262.30	20
5.41	0	250.10	12
6.25	0	257.42	6
7.08	0	253.76	8
7.91	0	247.66	8
8.33	0	244.00	0
9.50	0.32	235.46	6.5
10.83	0.72	261.08	16
11.66	0.80	274.50	18
12.41	0.96	272.06	21
13.24	0.88	271.45	17
13.99	0.80	265.96	14
14.99	0.88	274.50	16
16.24	1.28	286.70	25
17.58	1.28	283.04	20
18.49	2.16	270.84	33
20.33	2.24	267.79	30
21.07	2.80	247.05	34

Table B.1: Rotation velocity in the radial range from 3 to 21 kpc obtained from observation of HI and molecular clouds within the solar circle and HII, planetary nebulae and stars at the outer Galaxy. Data adapted from [162]. Data are shown in Figure 3.5a.

r [kpc]	σ_r [kpc]	v [km s^{-1}]	σ_v [km s^{-1}]
0.10	0.0	246.15	2.5
0.13	0.0	266.81	4.0
0.15	0.0	283.83	3.0
0.24	0.0	299.43	3.0
0.26	0.0	310.96	2.0
0.27	0.0	312.74	2.5
0.32	0.0	302.83	3.0
0.39	0.0	305.07	3.0
0.43	0.0	306.91	2.5
0.49	0.0	306.01	2.5
0.54	0.0	302.57	4.0
0.57	0.0	297.24	2.0
0.59	0.0	292.30	2.0
0.69	0.0	279.81	2.0
0.72	0.0	279.70	4.0
0.93	0.0	271.41	3.0
0.98	0.0	269.19	4.0
1.03	0.0	264.78	4.0
1.08	0.0	262.80	2.0
1.11	0.0	259.42	2.0
1.13	0.0	262.66	3.0
1.18	0.0	264.47	3.0
1.22	0.0	265.42	4.0
1.25	0.0	263.87	3.0
1.27	0.0	259.18	2.0
1.37	0.0	241.20	3.0
1.38	0.0	261.42	3.0
1.42	0.0	254.48	2.0
1.52	0.0	241.02	9.0
1.57	0.0	253.56	2.5
1.59	0.0	255.21	3.0
1.62	0.0	250.49	3.0
1.67	0.0	246.20	3.0
1.69	0.0	238.09	3.0
1.71	0.0	241.66	7.0
1.76	0.0	244.45	3.0
1.88	0.0	246.42	3.0
1.98	0.0	244.30	3.0
2.09	0.0	243.38	3.0
2.16	0.0	238.30	2.0
2.31	0.0	236.01	3.0
2.35	0.0	232.23	2.0
2.40	0.0	231.11	3.0
2.50	0.0	230.95	3.0

Table B.2: Rotation velocities in the radial range from 100 pc to 2.5 kpc from the observation of HI and CO in the inner Galaxy [93]. Data shown in Figure 3.5a.

r [kpc]	σ_r [kpc]	v [km s^{-1}]	σ_v [km s^{-1}]
1.92	0.0	245.69	7.0
2.02	0.0	241.75	1.5
2.16	0.0	233.42	1.5
2.45	0.0	229.50	3.0
2.53	0.0	236.59	3.0
2.55	0.0	231.90	1.5
2.60	0.0	230.90	2.0
2.65	0.0	232.59	2.0
2.69	0.0	228.79	1.5
2.74	0.0	231.21	1.5
2.79	0.0	231.55	1.5
2.84	0.0	230.43	1.5
2.89	0.0	230.53	1.5
2.94	0.0	234.53	2.0
2.99	0.0	231.59	2.0
3.04	0.0	234.74	1.5
3.06	0.0	236.14	1.5
3.10	0.0	226.76	1.5
3.16	0.0	232.32	1.5
3.19	0.0	235.89	1.5
3.23	0.0	233.55	1.5
3.29	0.0	233.14	1.5
3.33	0.0	228.99	1.5
3.35	0.0	233.57	1.5
3.38	0.0	239.95	1.5
3.44	0.0	245.76	1.5
3.47	0.0	240.43	1.5
3.51	0.0	236.41	1.5
3.55	0.0	238.49	1.5
3.59	0.0	241.18	1.5
3.63	0.0	241.19	2.0
3.67	0.0	243.00	2.0
3.72	0.0	240.41	2.0
3.76	0.0	239.69	1.5
3.82	0.0	238.78	2.0
3.84	0.0	241.90	1.5
3.90	0.0	241.12	1.5
3.94	0.0	245.27	1.5
3.98	0.0	246.38	1.5
4.02	0.0	247.48	1.5
4.06	0.0	241.88	1.5
4.12	0.0	245.61	1.5
4.14	0.0	253.24	1.5
4.17	0.0	256.90	1.5
4.21	0.0	258.98	1.5
4.26	0.0	259.33	1.5
4.31	0.0	261.26	1.5
4.36	0.0	262.21	1.5
4.41	0.0	260.12	1.5
4.44	0.0	252.35	1.2
4.49	0.0	252.57	1.5
4.53	0.0	254.53	1.5
4.57	0.0	253.80	2.0
4.61	0.0	267.35	1.5
4.64	0.0	262.75	3.5
4.66	0.0	269.99	3.0
4.70	0.0	259.50	1.5
4.75	0.0	245.94	1.5
4.80	0.0	244.09	2.0
4.85	0.0	246.87	2.0
4.90	0.0	243.07	1.5
4.95	0.0	242.92	1.5
4.99	0.0	254.00	1.5
5.04	0.0	252.43	1.5
5.08	0.0	252.92	1.5
5.10	0.0	260.06	1.5
5.14	0.0	256.41	2.0
5.16	0.0	258.88	1.5
5.19	0.0	259.29	1.5
5.24	0.0	257.92	1.5
5.29	0.0	260.46	1.5
5.34	0.0	258.37	2.0
5.39	0.0	259.69	1.5
5.43	0.0	252.74	1.5
5.49	0.0	257.57	1.5
5.52	0.0	255.17	1.5
5.56	0.0	255.54	1.5
5.59	0.0	253.26	2.0
5.64	0.0	253.12	1.5
5.68	0.0	248.09	1.5
5.73	0.0	250.75	1.5
5.76	0.0	251.89	1.5
5.80	0.0	257.51	1.5
5.84	0.0	256.17	1.5
5.90	0.0	258.68	2.0
5.93	0.0	255.91	1.5
5.98	0.0	256.14	1.5
6.03	0.0	255.02	1.5
6.10	0.0	255.67	1.5
6.15	0.0	249.89	3.0
6.17	0.0	254.22	1.5
6.22	0.0	256.27	1.5
6.27	0.0	261.25	1.5
6.30	0.0	268.98	1.5
6.36	0.0	271.73	1.5
6.40	0.0	248.44	1.5
6.45	0.0	249.88	1.5
6.47	0.0	266.78	1.5
6.52	0.0	267.00	1.5
6.57	0.0	269.30	1.5
6.61	0.0	273.30	1.5
6.62	0.0	274.13	1.5
6.66	0.0	277.79	3.5
6.71	0.0	269.48	1.5
6.76	0.0	250.06	2.0
6.81	0.0	253.33	1.5
6.86	0.0	253.55	15.0
6.91	0.0	255.48	2.0
6.96	0.0	254.97	2.0
7.01	0.0	253.98	2.0
7.04	0.0	256.33	1.5
7.06	0.0	254.56	1.5
7.10	0.0	262.26	2.0
7.13	0.0	246.53	1.5
7.17	0.0	251.05	1.5
7.20	0.0	250.60	1.5
7.25	0.0	256.31	1.5
7.30	0.0	246.04	1.5
7.35	0.0	257.36	1.5
7.40	0.0	253.07	2.0
7.45	0.0	254.76	2.0
7.50	0.0	251.32	2.0
7.52	0.0	253.22	1.2
7.57	0.0	247.22	1.6
7.62	0.0	251.31	1.5
7.64	0.0	246.25	1.5
7.66	0.0	244.73	1.5
7.74	0.0	250.60	4.0
7.78	0.0	250.49	1.5
7.82	0.0	255.25	3.0
7.87	0.0	249.99	1.2
7.88	0.0	250.44	1.2
7.94	0.0	244.29	1.3
7.99	0.0	244.52	1.5
8.04	0.0	247.30	1.5
8.07	0.0	243.77	1.3
8.09	0.0	243.47	1.5
8.13	0.0	247.02	1.5
8.18	0.0	258.10	1.5
8.23	0.0	256.00	1.5
8.28	0.0	257.56	1.5
8.33	0.0	253.15	1.5

Table B.3: Rotation velocities from ∼ 2 kpc to the Solar circle [94]. Data shown in Figure 3.5a.

r [kpc]	σ_r [kpc]	v [km s^{-1}]	σ_v [km s^{-1}]
10.75	0.0	185.65	7.01
10.57	0.0	241.26	11.85
10.82	0.0	189.64	6.25
11.36	0.0	217.98	2.53
10.81	0.0	164.85	10.27
13.04	0.0	126.17	2.61
10.85	0.0	237.57	13.02
11.76	0.0	257.36	13.27
11.86	0.0	149.99	2.67
11.50	0.0	240.46	5.5
14.35	0.0	217.53	14.21
12.07	0.0	253.14	7.26
12.68	0.0	298.05	4.15
10.57	0.0	213.32	11.78
14.17	0.0	294.73	5.92
13.66	0.0	279.48	12.48
13.71	0.0	180.72	11.31
12.65	0.0	89.46	6.73
12.78	0.0	236.61	6.65
11.66	0.0	250.87	9.82
13.12	0.0	233.30	8.63
15.23	0.0	160.53	10.77
11.58	0.0	229.37	11.6
10.79	0.0	205.31	5.15
11.42	0.0	222.03	11.32
10.88	0.0	223.21	17.38
10.67	0.0	173.29	6.03
11.01	0.0	203.62	8.06
9.89	0.0	178.13	6.05
9.73	0.0	195.20	2.01
9.83	0.0	212.68	5.05
10.91	0.0	204.89	4.06
10.85	0.0	264.96	6.09
10.01	0.0	175.77	6.09
10.68	0.0	224.86	13.54

Table B.4: Rotation velocities in the radial range from 9 to 15 kpc from the observation of carbon stars [99]. Data shown in Figure 3.5a.

r [kpc]	σ_r [kpc]	v [km s^{-1}]	σ_v [km s^{-1}]	r [kpc]	σ_r [kpc]	v [km s^{-1}]	σ_v [km s^{-1}]
2.16	0.0	245.30	0.0	8.33	0.0	244.09	0.0
2.30	0.0	238.42	0.0	8.32	0.0	244.10	0.0
2.44	0.0	232.50	0.0	8.32	0.0	244.64	0.0
2.57	0.0	233.15	0.0	8.31	0.0	244.02	0.0
2.71	0.0	231.70	0.0	8.30	0.0	247.71	0.0
2.85	0.0	233.88	0.0	8.28	0.0	246.57	0.0
2.99	0.0	235.06	0.0	8.27	0.0	246.57	0.0
3.12	0.0	232.07	0.0	8.25	0.0	248.82	0.0
3.25	0.0	235.64	0.0	8.23	0.0	246.61	0.0
3.39	0.0	265.77	0.0	8.20	0.0	247.13	0.0
3.52	0.0	264.53	0.0	8.18	0.0	247.20	0.0
3.65	0.0	244.21	0.0	8.15	0.0	249.41	0.0
3.78	0.0	240.58	0.0	8.12	0.0	250.80	0.0
3.91	0.0	243.51	0.0	8.08	0.0	252.12	0.0
4.04	0.0	247.49	0.0	8.05	0.0	253.74	0.0
4.17	0.0	254.61	0.0	8.01	0.0	255.90	0.0
4.29	0.0	287.08	0.0	7.97	0.0	252.74	0.0
4.41	0.0	259.84	0.0	7.92	0.0	251.46	0.0
4.54	0.0	259.53	0.0	7.88	0.0	251.20	0.0
4.66	0.0	261.37	0.0	7.83	0.0	241.49	0.0
4.78	0.0	257.80	0.0	7.78	0.0	250.12	0.0
4.90	0.0	252.00	0.0	7.72	0.0	252.10	0.0
5.01	0.0	255.42	0.0	7.67	0.0	252.30	1.0
5.13	0.0	257.58	0.0	7.55	0.0	251.88	0.0
5.24	0.0	258.96	0.0	7.49	0.0	254.44	0.0
5.35	0.0	251.15	0.0	7.42	0.0	257.91	0.0
5.46	0.0	253.04	0.0	7.36	0.0	256.55	0.0
5.57	0.0	251.23	0.0	7.29	0.0	250.86	0.0
5.68	0.0	256.93	0.0	7.21	0.0	254.50	0.0
5.79	0.0	254.29	0.0	7.14	0.0	257.22	0.0
5.89	0.0	254.76	0.0	7.06	0.0	257.68	0.0
5.99	0.0	256.28	0.0	6.99	0.0	260.15	0.0
6.09	0.0	260.92	0.0	6.91	0.0	257.80	0.0
6.29	0.0	265.52	0.0	6.82	0.0	254.65	0.0
6.38	0.0	266.70	0.0	6.74	0.0	247.42	0.0
6.47	0.0	267.34	0.0	6.65	0.0	254.16	0.0
6.56	0.0	269.26	0.0	6.56	0.0	260.59	0.0
6.65	0.0	271.48	0.0	6.47	0.0	260.38	0.0
6.74	0.0	263.65	0.0	6.38	0.0	260.12	0.0
6.82	0.0	255.51	0.0	6.29	0.0	260.16	0.0
6.91	0.0	257.67	0.0	6.19	0.0	255.38	0.0
6.99	0.0	253.44	0.0	6.09	0.0	251.65	0.0
7.06	0.0	255.60	0.0	5.99	0.0	250.18	0.0
7.14	0.0	256.48	0.0	5.89	0.0	248.79	0.0
7.21	0.0	255.47	0.0	5.79	0.0	243.55	0.0
7.29	0.0	254.28	0.0	5.68	0.0	241.07	0.0
7.36	0.0	252.89	0.0	5.57	0.0	238.91	0.0
7.42	0.0	267.30	0.0	5.46	0.0	245.11	0.0
7.49	0.0	250.54	0.0	5.35	0.0	245.29	0.0
7.55	0.0	251.52	0.0	5.24	0.0	247.49	0.0
7.61	0.0	247.67	0.0	5.13	0.0	247.21	0.0
7.67	0.0	254.01	0.0	5.01	0.0	248.96	0.0
7.72	0.0	244.78	0.0	4.90	0.0	244.68	0.0
7.78	0.0	247.80	0.0	4.78	0.0	249.39	0.0
7.83	0.0	246.49	0.0	4.66	0.0	248.56	0.0
7.88	0.0	245.10	0.0	4.54	0.0	250.50	0.0
7.92	0.0	246.09	0.0	4.41	0.0	255.08	0.0
7.97	0.0	247.12	0.0	4.29	0.0	280.73	0.0
8.01	0.0	246.50	0.0	4.17	0.0	273.89	0.0
8.05	0.0	245.57	0.0	4.04	0.0	273.84	0.0
8.08	0.0	248.46	0.0	3.91	0.0	268.15	0.0
8.12	0.0	248.97	0.0	3.78	0.0	245.46	0.0
8.15	0.0	254.77	0.0	3.65	0.0	244.09	0.0
8.18	0.0	258.30	0.0	3.52	0.0	246.84	0.0
8.20	0.0	255.18	0.0	3.39	0.0	251.50	0.0
8.23	0.0	256.98	0.0	3.25	0.0	247.96	0.0
8.25	0.0	260.05	0.0	3.12	0.0	241.47	1.0
8.27	0.0	275.12	0.0	2.99	0.0	242.99	0.0
8.28	0.0	287.32	0.0	2.85	0.0	240.10	0.0
8.30	0.0	255.03	0.0	2.71	0.0	242.68	0.0
8.31	0.0	256.70	0.0	2.57	0.0	243.76	0.0
8.32	0.0	254.63	0.0	2.44	0.0	254.22	0.0
8.32	0.0	258.37	0.0	2.30	0.0	249.65	0.0
8.33	0.0	255.80	0.0	2.16	0.0	258.72	0.0

Table B.5: Rotation velocities in the radial range from about 2 kpc to about 8 kpc [95], shown in Figure 3.5a.

r [kpc]	σ_r [kpc]	v [km s^{-1}]	σ_v [km s^{-1}]
6.58	0.29	220.50	12.41
6.54	0.30	221.25	10.89
6.47	0.20	299.18	15.30
8.15	0.05	291.28	21.99
6.15	0.19	249.59	9.04
6.35	0.66	248.94	16.83
5.50	0.81	287.21	20.80
6.40	0.18	267.20	4.79
5.44	0.80	245.33	20.17
5.51	0.73	246.28	18.22
6.61	0.47	266.24	12.12
7.71	0.20	259.36	5.09
7.53	0.06	254.71	2.20
7.44	0.07	245.07	2.07
7.50	0.04	245.75	1.82
7.70	0.14	251.25	3.72
9.57	1.44	246.28	35.98
7.90	0.01	248.25	0.43
8.39	0.40	245.69	10.21
8.38	0.11	240.52	3.36
8.29	0.01	242.94	3.01
8.32	0.01	248.12	1.52
9.73	0.72	198.22	18.10
8.96	0.23	205.34	5.75
8.47	0.03	258.22	1.28
8.43	0.04	246.55	1.04
11.87	1.65	181.97	41.22
11.19	1.55	207.73	38.81
8.41	0.03	228.74	0.98
10.24	0.91	225.30	22.91
8.60	0.10	231.04	3.59
8.81	0.16	230.48	4.07
8.51	0.07	235.99	2.20
9.84	0.64	218.51	15.95
8.65	0.04	242.01	1.43
9.97	0.18	229.38	4.55
11.42	1.23	428.20	30.70
8.92	0.19	245.25	4.80
12.33	1.55	263.04	38.70
9.59	0.24	216.01	6.25
9.78	0.55	200.15	13.79
9.98	0.76	204.30	18.96
9.80	0.55	206.72	13.85
10.28	0.73	228.35	18.18
9.54	0.42	211.06	11.28
9.15	0.28	219.36	7.09
10.66	0.85	241.92	21.38
9.67	0.45	213.47	11.36
10.01	0.62	235.25	15.83
9.35	0.32	195.12	8.02
9.91	0.55	232.66	13.82
9.77	0.49	233.96	12.31
9.97	0.16	199.04	8.35
10.07	0.25	269.42	6.95
8.98	0.24	239.61	6.78
11.45	0.75	257.99	18.72
15.74	0.77	316.05	19.36
16.35	2.33	216.07	58.43
14.13	1.74	231.97	43.59
14.25	0.58	240.3	14.57
13.53	0.78	281.95	19.76
12.52	0.58	239.12	14.97
8.72	0.038	282.69	9.16
11.93	1.09	268.38	30.35
12.84	0.40	248.10	10.18
16.87	2.58	241.92	64.64
8.78	0.14	255.71	5.35
8.73	0.13	202.11	3.58
12.18	0.98	246.41	24.65
13.18	0.49	256.46	12.37
9.10	0.14	242.93	4.32
9.84	0.19	242.93	4.62
8.79	0.04	234.92	3.46
17.17	2.81	260.14	70.44
13.24	0.76	222.49	18.91
14.85	0.67	257.88	16.74
11.14	0.71	254.51	17.79
10.94	1.06	175.32	26.40
9.23	0.11	234.54	3.08
12.23	1.32	273.66	33.04
11.77	0.50	239.02	12.54
12.98	1.78	251.12	19.46
9.89	0.52	259.42	13.01
12.33	1.17	262.43	29.36
9.84	0.36	206.26	9.09
12.54	0.67	268.36	16.94
9.22	0.31	235	7.91
10.94	0.44	228.90	11.15
16.75	2.71	389.42	67.63

Table B.6: Rotation velocities in the radial range of 8 to 17 kpc [96], shown in Figure 3.5a.

r [kpc]	σ_r [kpc]	v [km s^{-1}]	σ_v [km s^{-1}]
2.25	0.15	229.77	8
2.74	0.15	222	5
3.14	0.15	229.77	10
3.72	0.15	236.43	8
4.12	0.15	255.3	13
4.70	0.15	244.2	4
5.10	0.15	241.98	4
5.68	0.15	239.76	5
6.08	0.15	249.75	5
6.57	1.15	255.3	5
7.06	1.15	251.97	2
7.64	0.15	249.75	4
8.04	0.15	246.42	10
8.53	0.15	245.31	5
9.02	0.15	238.65	12
9.60	0.15	233.1	21
9.99	0.15	227.55	13
10.58	0.15	230.88	12
11.07	0.15	230.88	13
12.25	0.25	255.3	22
13.52	0.7	255.3	15
15.48	0.5	310.8	32

Table B.7: Rotation velocities in the radial range from about 2 to about 15 kpc [98], shown in Figure 3.5a.

Nr.	r_{min} [kpc]	r_{max} [kpc]	r [kpc]	σ_r [kpc]	v [km s^{-1}]	σ_v [km s^{-1}]
1	0.0	1.0	0.0	0.0	0.0	0.0
2	1.0	1.5	0.0	0.0	0.0	0.0
3	1.5	2.0	0.0	0.0	0.0	0.0
4	2.0	2.5	0.0	0.0	0.0	0.0
5	2.5	3.0	2.92	0.01	239.12	16
6	3.0	4.0	3.75	0.01	253.76	15
7	4.0	5.0	4.58	0.01	262.30	20
8	5.0	6.0	5.41	0.02	250.10	12
9	6.0	7.0	6.25	0.02	257.42	6
10	7.0	8.0	7.45	0.02	250.71	5.66
11	8.0	8.5	8.33	0.03	244.00	0.77
12	8.5	9.0	0.0	0.0	0.0	0.0
13	9.0	10.0	9.50	0.32	235.46	6.5
14	10.0	11.0	10.83	0.72	261.08	16
15	11.0	12.0	11.66	0.80	274.50	18
16	12.0	14.0	13.31	0.50	268.99	9.61
17	14.0	16.0	14.99	0.88	274.50	16
18	16.0	22.0	17.79	0.75	275.40	11.96

Table B.8: Averaged values of the Galactocentric distances and the rotation velocities given in Table B.1 in 18 radial bins. The averaged velocities are shown in Figure 3.5b.

Nr.	r_{min} [kpc]	r_{max} [kpc]	r [kpc]	σ_r [kpc]	v [km s^{-1}]	σ_v [km s^{-1}]
1	0.0	1.0	0.15	0.00	281.42	0.84
2	1.0	1.5	1.20	0.00	259.35	0.77
3	1.5	2.0	1.68	0.00	247.45	1.02
4	2.0	2.5	2.28	0.00	235.31	1.03
5	2.5	3.0	0.0	0.0	0.0	0.0
6	3.0	4.0	0.0	0.0	0.0	0.0
7	4.0	5.0	0.0	0.0	0.0	0.0
8	5.0	6.0	0.0	0.0	0.0	0.0
9	6.0	7.0	0.0	0.0	0.0	0.0
10	7.0	8.0	0.0	0.0	0.0	0.0
11	8.0	8.5	0.0	0.0	0.0	0.0
12	8.5	9.0	0.0	0.0	0.0	0.0
13	9.0	10.0	0.0	0.0	0.0	0.0
14	10.0	11.0	0.0	0.0	0.0	0.0
15	11.0	12.0	0.0	0.0	0.0	0.0
16	12.0	14.0	0.0	0.0	0.0	0.0
17	14.0	16.0	0.0	0.0	0.0	0.0
18	16.0	22.0	0.0	0.0	0.0	0.0

Table B.9: Averaged values of the Galactocentric distances and the rotation velocities given in Table B.2 in 18 radial bins. The averaged velocities are shown in Figure 3.5b.

Nr.	r_{min} [kpc]	r_{max} [kpc]	r [kpc]	σ_r [kpc]	v [km s^{-1}]	σ_v [km s^{-1}]
1	0.0	1.0	0.0	0.0	0.0	0.0
2	1.0	1.5	0.0	0.0	0.0	0.0
3	1.5	2.0	1.92	0.01	245.69	7.0
4	2.0	2.5	2.18	0.00	236.68	1.0
5	2.5	3.0	2.73	0.00	231.35	0.51
6	3.0	4.0	3.45	0.00	237.91	0.32
7	4.0	5.0	4.46	0.00	253.46	0.32
8	5.0	6.0	5.46	0.00	255.72	0.32
9	6.0	7.0	6.47	0.00	261.35	0.35
10	7.0	8.0	7.45	0.00	250.75	0.31
11	8	8.5	8.17	0.01	250.52	0.52
12	8.5	9.0	0.0	0.0	0.0	0.0
13	9.0	10.0	0.0	0.0	0.0	0.0
14	10.0	11.0	0.0	0.0	0.0	0.0
15	11.0	12.0	0.0	0.0	0.0	0.0
16	12.0	14.0	0.0	0.0	0.0	0.0
17	14.0	16.0	0.0	0.0	0.0	0.0
18	16.0	22.0	0.0	0.0	0.0	0.0

Table B.10: Averaged values of the Galactocentric distances and the rotation velocities given in Table B.3 in 18 radial bins. The averaged velocities are shown in Figure 3.5b.

Nr.	r_{min} [kpc]	r_{max} [kpc]	r [kpc]	σ_r [kpc]	v [km s^{-1}]	σ_v [km s^{-1}]
1	0.0	1.0	0.0	0.0	0.0	0.0
2	1.0	1.5	0.0	0.0	0.0	0.0
3	1.5	2.0	0.0	0.0	0.0	0.0
4	2.0	2.5	0.0	0.0	0.0	0.0
5	2.5	3.0	0.0	0.0	0.0	0.0
6	3.0	4.0	0.0	0.0	0.0	0.0
7	4.0	5.0	0.0	0.0	0.0	0.0
8	5.0	6.0	0.0	0.0	0.0	0.0
9	6.0	7.0	0.0	0.0	0.0	0.0
10	7.0	8.0	0.0	0.0	0.0	0.0
11	8.0	8.5	0.0	0.0	0.0	0.0
12	8.5	9.0	0.0	0.0	0.0	0.0
13	9.0	10.0	9.81	0.02	195.90	1.78
14	10.0	11.0	10.69	0.00	203.00	1.95
15	11.0	12.0	11.51	0.01	195.83	1.63
16	12.0	14.0	12.92	0.01	183.31	1.84
17	14.0	16.0	14.56	0.03	258.17	4.87
18	16.0	22.0	0.0	0.0	0.0	0.0

Table B.11: Averaged values of the Galactocentric distances and the rotation velocities given in Table B.4 in 18 radial bins. The averaged velocities are shown in Figure 3.5b.

Nr.	r_{min} [kpc]	r_{max} [kpc]	r [kpc]	σ_r [kpc]	v [km s^{-1}]	σ_v [km s^{-1}]
1	0.0	1.0	0.0	0.0	0.0	0.0
2	1.0	1.5	0.0	0.0	0.0	0.0
3	1.5	2.0	0.0	0.0	0.0	0.0
4	2.0	2.5	2.28	0.00	245.81	0.32
5	2.5	3.0	2.76	0.00	237.73	0.27
6	3.0	4.0	3.48	0.00	247.12	0.21
7	4.0	5.0	4.44	0.00	258.69	0.20
8	5.0	6.0	5.48	0.00	250.06	0.18
9	6.0	7.0	6.55	0.00	259.44	0.18
10	7.0	8.0	7.53	0.00	251.94	0.14
11	8.0	8.5	8.21	0.00	252.16	0.14
12	8.5	9.0	0.0	0.0	0.0	0.0
13	9.0	10.0	0.0	0.0	0.0	0.0
14	10.0	11.0	0.0	0.0	0.0	0.0
15	11.0	12.0	0.0	0.0	0.0	0.0
16	12.0	14.0	0.0	0.0	0.0	0.0
17	14.0	16.0	0.0	0.0	0.0	0.0
18	16.0	22.0	0.0	0.0	0.0	0.0

Table B.12: Averaged values of the Galactocentric distances and the rotation velocities given in Table B.5 in 18 radial bins. The averaged velocities are shown in Figure 3.5b.

Nr.	r_{min} [kpc]	r_{max} [kpc]	r [kpc]	σ_r [kpc]	v [km s^{-1}]	σ_v [km s^{-1}]
1	0.0	1.0	0.0	0.0	0.0	0.0
2	1.0	1.5	0.0	0.0	0.0	0.0
3	1.5	2.0	0.0	0.0	0.0	0.0
4	2.0	2.5	0.0	0.0	0.0	0.0
5	2.5	3.0	0.0	0.0	0.0	0.0
6	3.0	4.0	0.0	0.0	0.0	0.0
7	4.0	5.0	0.0	0.0	0.0	0.0
8	5.0	6.0	5.49	0.45	258.13	11.33
9	6.0	7.0	6.39	0.09	257.35	3.42
10	7.0	8.0	7.90	0.00	248.33	0.40
11	8.0	8.5	8.30	0.00	242.93	0.56
12	8.5	9.0	8.70	0.02	235.95	0.91
13	9.0	10.0	9.51	0.06	228.64	1.56
14	10.0	11.0	10.29	0.19	248.46	4.92
15	11.0	12.0	11.57	0.33	244.50	8.24
16	12.0	14.0	12.84	0.20	252.55	5.19
17	14.0	16.0	14.49	0.42	246.92	10.66
18	16.0	22.0	16.74	1.47	236.61	36.92

Table B.13: Averaged values of the Galactocentric distances and the rotation velocities given in Table B.6 in 18 radial bins. The averaged velocities are shown in Figure 3.5b.

Nr.	r_{min} [kpc]	r_{max} [kpc]	r [kpc]	σ_r [kpc]	v [km s^{-1}]	σ_v [km s^{-1}]
1	0.0	1.0	0.0	0.0	0.0	0.0
2	1.0	1.5	0.0	0.0	0.0	0.0
3	1.5	2.0	0.0	0.0	0.0	0.0
4	2.0	2.5	2.25	0.15	229.77	8.0
5	2.5	3.0	2.74	0.15	222.00	5.0
6	3.0	4.0	3.43	0.11	233.83	6.25
7	4.0	5.0	4.41	0.11	245.16	3.82
8	5.0	6.0	5.39	0.11	241.11	3.12
9	6.0	7.0	6.32	0.11	252.53	3.54
10	7.0	8.0	7.35	0.11	251.53	1.79
11	8.0	8.5	8.04	0.15	246.42	10.0
12	8.5	9.0	8.53	0.15	245.31	5.0
13	9.0	10.0	9.54	0.09	233.48	8.13
14	10.0	11.0	10.58	0.15	230.88	12.0
15	11.0	12.0	11.07	0.15	230.88	13.0
16	12.0	14.0	12.39	0.24	255.30	12.39
17	14.0	16.0	0.0	0.0	0.0	0.0
18	16.0	22.0	0.0	0.0	0.0	0.0

Table B.14: Averaged values of the Galactocentric distances and the rotation velocities given in Table B.7 in 18 radial bins. The averaged velocities are shown in Figure 3.5b.

C
Single Profile plots

In this appendix the results of the Single Profile (SP) density model for the Navarro-Frenk-White (NFW) profile, the Binney-Evans (BE) profile, the Moore profile, the pseudo-isothermal (PISO) profile and the 240 profile are presented. For each profile the energy spectrum and the longitudinal distribution of the diffuse Galactic gamma rays measured with EGRET, the DM density distribution in the Galactic plane, the resulting Galactic mass, the rotation curve of the Galactic disc and the vertical gravitational potential at the position of the Sun are shown. In general, the profiles result in too high ring densities, which yield a local DM density in disagreement with the local DM density given by the Oort limit and the local surface density (see too large vertical potential in the following figures).

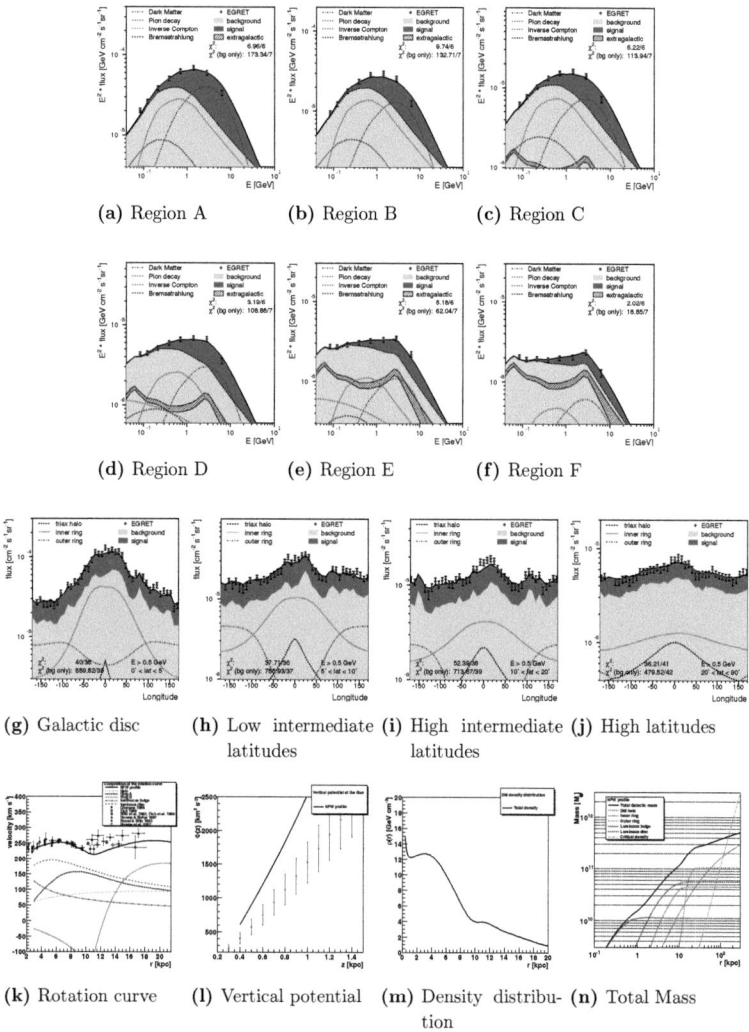

Figure C.1: Fit results for the NFW halo profile. The energy spectra of the diffuse Galactic gamma radiation are shown in **(a)** – **(f)** for the different regions defined in Section 4.2.1. The longitudinal distribution of the diffuse gamma radiation above a photon energy of 500 MeV is shown in **(g)** – **(j)**. The rotation curve in the Galactic disc, the gravitational potential perpendicular to the Galactic plane at the position of the Sun, the density distribution in the disc and the total mass of the Galaxy are shown in **(k)** – **(n)**. The parameters of the density distribution are given in Table 4.2.

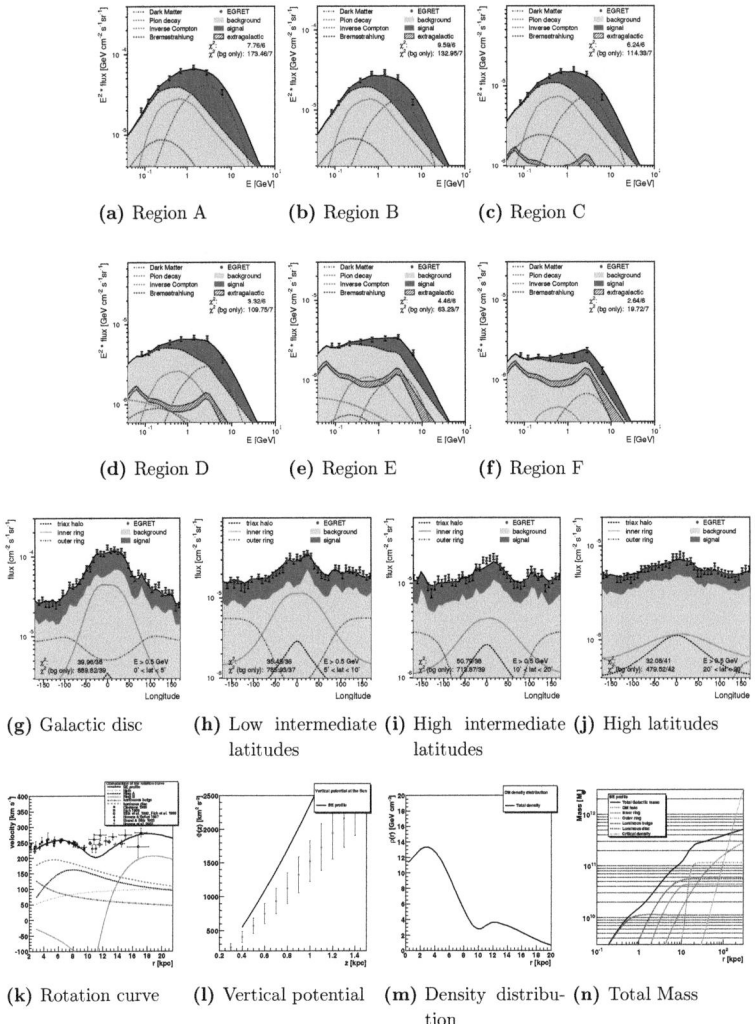

Figure C.2: Fit results for the BE halo profile. The energy spectra of the diffuse Galactic gamma radiation are shown in **(a)** – **(f)** for the different regions defined in Section 4.2.1. The longitudinal distribution of the diffuse gamma radiation above a photon energy of 500 MeV is shown in **(g)** – **(j)**. The rotation curve in the Galactic disc, the gravitational potential perpendicular to the Galactic plane at the position of the Sun, the density distribution in the disc and the total mass of the Galaxy are shown in **(k)** – **(n)**. The parameters of the density distribution are given in Table 4.2.

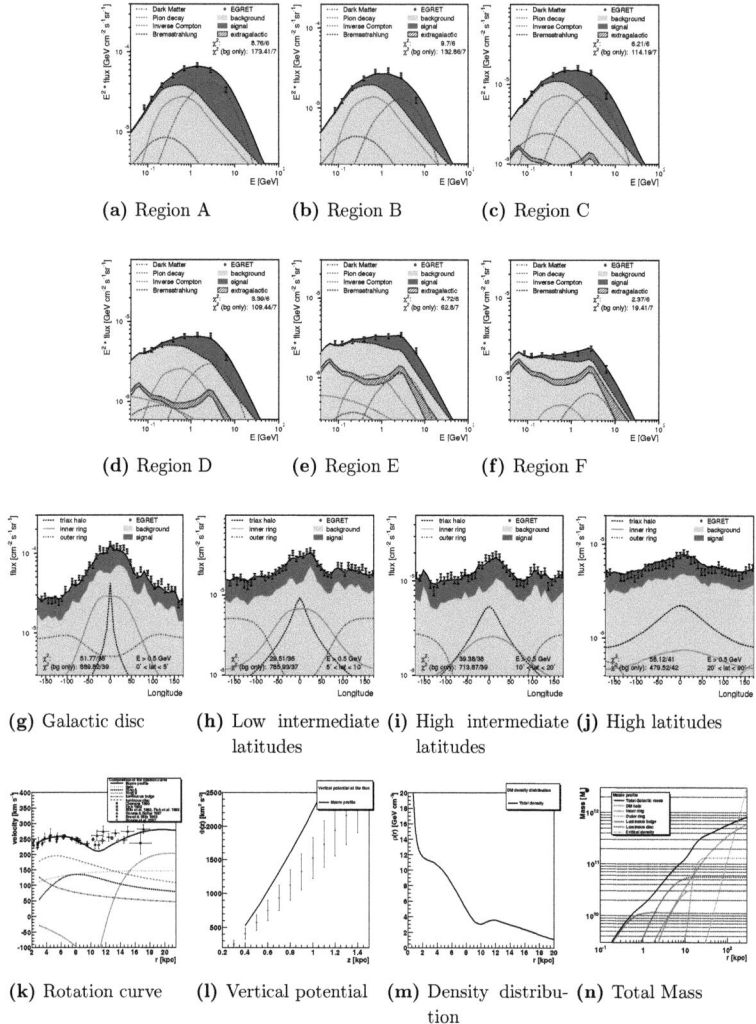

Figure C.3: Fit results for the Moore halo profile. The energy spectra of the diffuse Galactic gamma radiation are shown in **(a)** – **(f)** for the different regions defined in Section 4.2.1. The longitudinal distribution of the diffuse gamma radiation above a photon energy of 500 MeV is shown in **(g)** – **(j)**. The rotation curve in the Galactic disc, the gravitational potential perpendicular to the Galactic plane at the position of the Sun, the density distribution in the disc and the total mass of the Galaxy are shown in **(k)** – **(n)**. The parameters of the density distribution are given in Table 4.2.

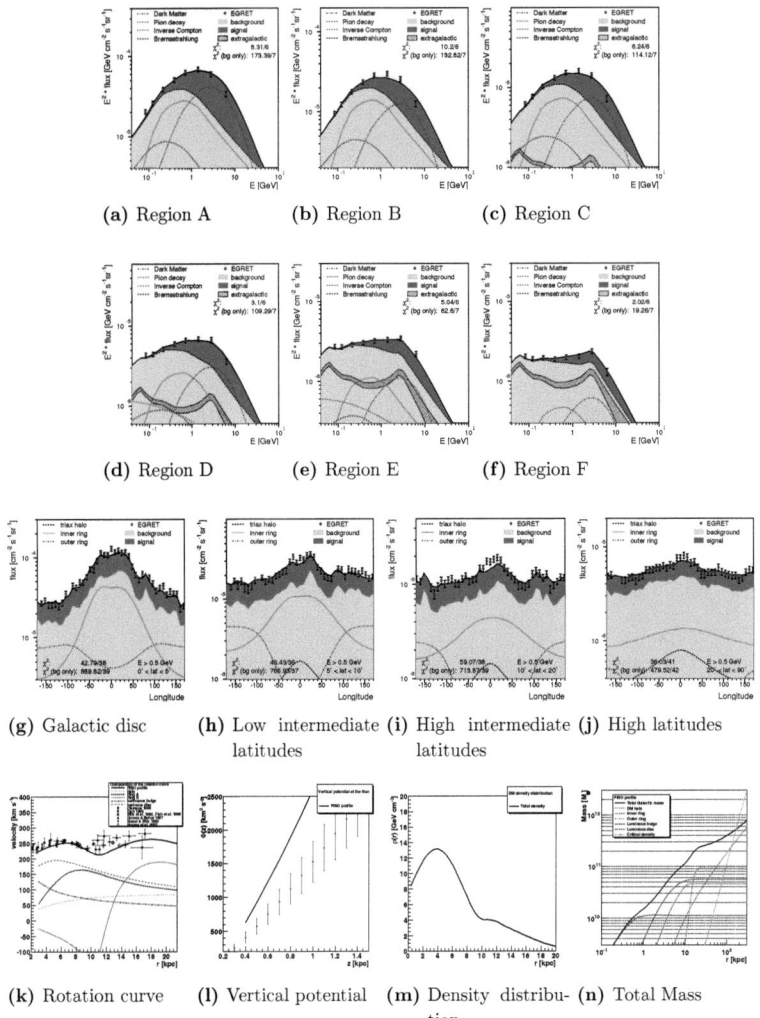

Figure C.4: Fit results for the PISO halo profile. The energy spectra of the diffuse Galactic gamma radiation are shown in (a) – (f) for the different regions defined in Section 4.2.1. The longitudinal distribution of the diffuse gamma radiation above a photon energy of 500 MeV is shown in (g) – (j). The rotation curve in the Galactic disc, the gravitational potential perpendicular to the Galactic plane at the position of the Sun, the density distribution in the disc and the total mass of the Galaxy are shown in (k) – (n). The parameters of the density distribution are given in Table 4.2.

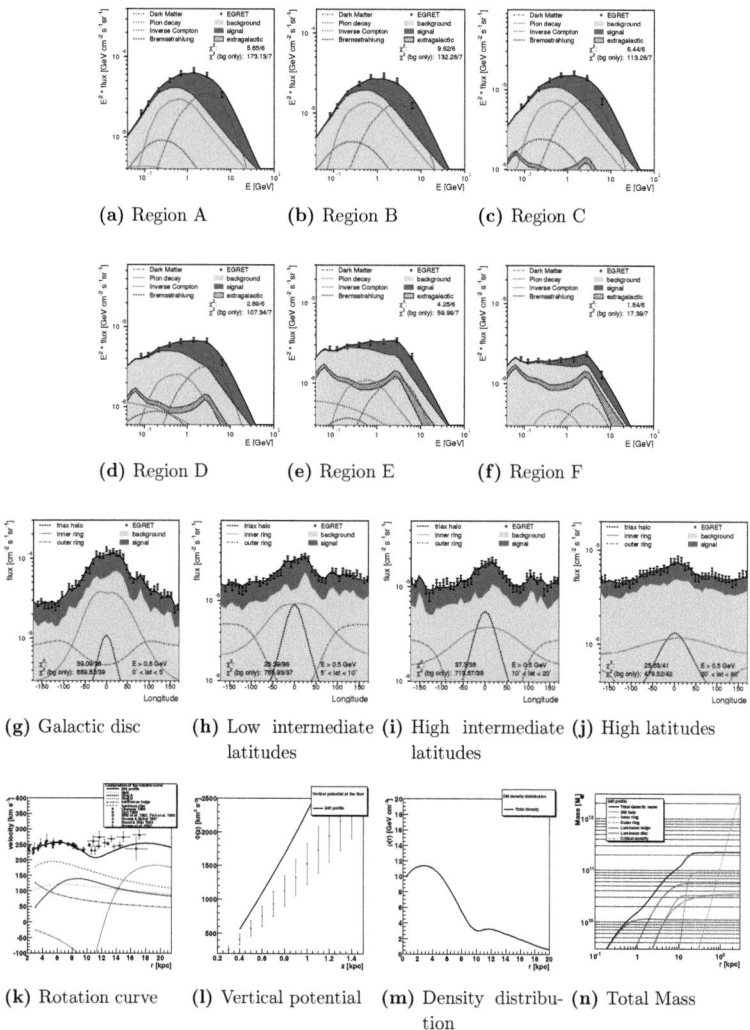

Figure C.5: Fit results for the 240 halo profile. The energy spectra of the diffuse Galactic gamma radiation are shown in **(a)** – **(f)** for the different regions defined in Section 4.2.1. The longitudinal distribution of the diffuse gamma radiation above a photon energy of 500 MeV is shown in **(g)** – **(j)**. The rotation curve in the Galactic disc, the gravitational potential perpendicular to the Galactic plane at the position of the Sun, the density distribution in the disc and the total mass of the Galaxy are shown in **(k)** – **(n)**. The parameters of the density distribution are given in Table 4.2.

D
Double Profile plots

In this appendix the results of the Double Profile (SP) density model for combinations the Navarro-Frenk-White (NFW) profile, the Binney-Evans (BE) profile, the Moore profile, the pseudo-isothermal (PISO) profile and the 240 profile are presented. In Section 4.2.2 it was shown that the constraint from the total Galactic mass and the local rotation velocity is not simultaneously fulfilled by the profile combinations with a Moore or a 240 profile for the diffuse DM component. Therefore, these profile combinations are not taken into account. The fit results of the remaining combinations are summarised in Figure 4.8. Here, for each profile combination the energy spectrum and the longitudinal distribution of the diffuse Galactic gamma rays measured with EGRET, the DM density distribution in the Galactic plane, the resulting Galactic mass, the rotation curve of the Galactic disc and the vertical gravitational potential at the position of the Sun are shown. The best results are obtained for profile combinations with a 240 profile for the clumpy DM component since this profile provides a higher DM density at intermediate latitudes (compare fit of the longitudinal gamma ray distribution at high intermediate latitudes for the different combinations).

Double Profile plots

Parameter	NFW-NFW	NFW-BE	NFW-Moore	NFW-PISO	BE-NFW	BE-BE	BE-Moore	BE-PISO	PISO-NFW	PISO-BE	PISO-Moore	PISO-240
$\rho_{\odot,\text{diff}}$ [GeV cm^{-3}]	0.304	0.304	0.304	0.304	0.348	0.348	0.348	0.348	0.167	0.167	0.167	0.167
r_{0} [kpc]	8.3	8.3	8.3	8.3	8.3	8.3	8.3	8.3	8.3	8.3	8.3	8.3
α_{diff}	1.0	1.0	1.0	1.0	1.0	1.0	1.0	1.0	2.0	2.0	2.0	2.0
β_{diff}	3.0	3.0	3.0	3.0	3.0	3.0	3.0	3.0	0.0	0.0	0.0	0.0
γ_{diff}	1.0	1.0	1.0	1.0	0.3	0.3	0.3	0.3	0.0	0.0	0.0	0.0
a_{diff} [kpc]	20.0	20.0	20.0	20.0	10.2	10.2	10.2	10.2	5.0	5.0	5.0	5.0
$\epsilon_{xy,\text{diff}}$	1.0	1.0	1.0	1.0	1.0	1.0	1.0	1.0	1.0	1.0	1.0	1.0
$\epsilon_{z,\text{diff}}$	1.0	1.0	1.0	1.0	1.0	1.0	1.0	1.0	1.0	1.0	1.0	1.0
$\rho_{\odot,\text{clump}}$ [GeV cm^{-3}]	0.021	0.01	0.018	0.018	0.017	0.019	0.022	0.014	0.035	0.04	0.025	0.018
α_{clump}	1.0	1.0	1.5	2.0	1.0	1.0	1.5	2.0	1.0	1.0	1.5	2.0
β_{clump}	3.0	3.0	3.0	2.0	3.0	3.0	3.0	2.0	3.0	3.0	3.0	2.0
γ_{clump}	1.0	0.3	1.5	0.0	1.0	0.3	1.5	0.0	1.0	0.3	1.5	0.0
a_{clump} [kpc]	20.0	30.0	30.0	5.0	20.0	30.0	30.0	5.0	30.0	30.0	30.0	5.0
$\epsilon_{xy,\text{clump}}$	0.85	0.85	0.85	0.85	0.85	0.85	0.85	0.85	0.85	0.85	0.85	0.85
$\epsilon_{z,\text{clump}}$	0.7	0.7	0.7	0.7	0.7	0.7	0.7	0.7	0.7	0.7	0.7	0.7
d_{lc} [°]	72.6	137.9	153.6	56.8	60.2	66.0	64.4	58.5	65.8	63.2	67.5	60.1
ρ_{IR} [GeV cm^{-3}]	3.35	3.40	3.24	4.37	3.12	3.18	3.03	4.03	9.61	8.48	9.09	9.39
R_{IR} [kpc]	3.23	3.12	3.37	2.99	3.26	3.48	3.32	3.10	2.56	2.51	2.49	3.13
$\sigma_{r,\text{IR}}$ [kpc]	2.91	2.92	2.86	2.76	2.95	2.92	2.85	2.76	3.48	3.56	3.48	3.48
$\sigma_{z,\text{IR}}$ [kpc]	0.35	0.39	0.34	0.39	0.32	0.33	0.34	0.37	0.41	0.50	0.40	0.41
$\epsilon_{xy,\text{IR}}$	0.85	0.85	0.85	0.85	0.85	0.85	0.85	0.85	0.85	0.85	0.85	0.85
ϕ_{IR} [°]	-119.46	-117.52	-121.75	-96.82	60.35	-117.64	-121.88	-93	-120.0	-113.96	-113.96	-121.05
M_{IR} [10^{9} M$_{\odot}$]	8.33	9.24	7.97	10.86	8.14	7.85	7.28	9.65	29.3	32.1	26.8	32.9
ρ_{OR} [GeV cm^{-3}]	0.68	0.71	0.53	0.99	0.87	0.66	0.49	0.78	2.87	2.62	1.82	2.63
R_{OR} [kpc]	12.90	12.64	13.02	12.91	12.80	12.94	12.99	12.85	12.53	12.51	12.58	13.03
$\sigma_{r,\text{OR}}$ [kpc]	3.98	3.73	6.05	4.39	2.27	3.73	5.99	4.40	2.34	3.73	4.77	4.09
$\sigma_{z,\text{OR}}$ [kpc]	4.0	4.0	4.0	4.0	4.0	4.0	4.0	4.0	4.0	4.0	4.0	4.0
$\epsilon_{xy,\text{OR}}$	0.88	0.86	1.03	0.95	0.75	0.87	1.03	0.70	0.93	0.71	0.88	0.88
ϕ_{OR} [°]	0.95	0.95	0.95	0.95	0.95	0.95	0.95	0.95	0.95	0.95	0.95	0.95
	-112.11	-113.65	-65.70	-112.48	-113.60	-109.95	-88.23	-32.64	-112.5	-166.61	-90.0
M_{OR} [10^{10} M$_{\odot}$]	1.98	1.87	2.72	3.38	1.35	1.77	2.51	2.60	4.14	5.60	5.51	7.87
boost factor	298.23	263.33	227.32	203.49	307.33	311.16	316.59	238.41	93.38	65.32	99.31	58.76
χ^{2}_{long} / 157	162.25	162.12	157.97	162.38	158.84	163.55	156.25	160.92	159.29	163.36	157.43	171.83
probability [%]	34.9	35.2	44.1	34.7	42.2	32.3	47.9	37.7	41.2	32.7	45.3	18.3
M_{tot} [10^{12} M$_{\odot}$]	1.11	1.09	1.11	1.18	1.09	1.09	1.11	1.13	1.28	1.27	1.29	1.36
$\rho_{\odot,\text{tot}}$ [M$_{\odot}$ pc^{-3}]	0.10	0.09	0.09	0.10	0.10	0.10	0.10	0.09	0.14	0.14	0.14	0.16
Σ_{tot} [M$_{\odot}$ pc^{-2}]	78.24	78.22	77.69	80.26	78.12	77.97	77.07	78.75	104.6	109.22	98.85	114.50

Table D.1: In Section 4.2.2 three of the fifteen considered halo profile combination, which fitted the data best, were considered. In this table the parameters of the twelve remaining profile combinations are presented.

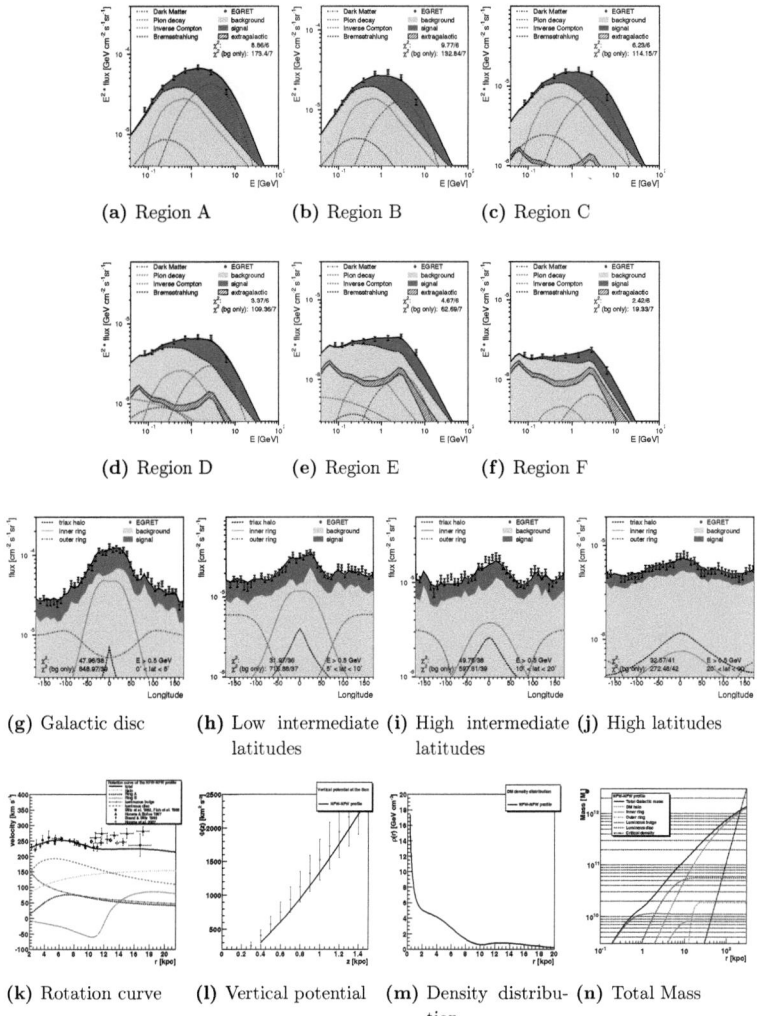

Figure D.1: Fit results for the NFW-NFW profile combination. The energy spectra of the diffuse Galactic gamma radiation are shown in **(a)** – **(f)** for the different regions defined in Section 4.2.1. The longitudinal distribution of the diffuse gamma radiation above a photon energy of 500 MeV is shown in **(g)** – **(j)**. The rotation curve in the Galactic disc, the gravitational potential perpendicular to the Galactic plane at the position of the Sun, the density distribution in the disc and the total mass of the Galaxy are shown in **(k)** – **(n)**. The parameters of the density distribution are given in Table D.1.

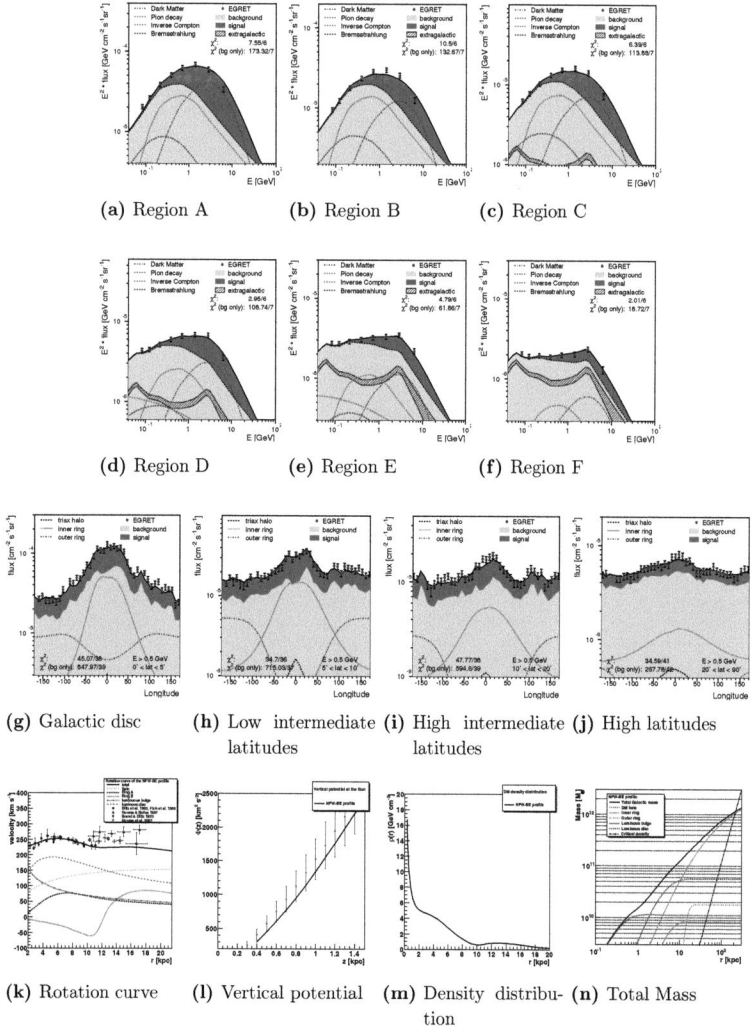

Figure D.2: Fit results for the NFW-BE profile combination. The energy spectra of the diffuse Galactic gamma radiation are shown in **(a) – (f)** for the different regions defined in Section 4.2.1. The longitudinal distribution of the diffuse gamma radiation above a photon energy of 500 MeV is shown in **(g) – (j)**. The rotation curve in the Galactic disc, the gravitational potential perpendicular to the Galactic plane at the position of the Sun, the density distribution in the disc and the total mass of the Galaxy are shown in **(k) – (n)**. The parameters of the density distribution are given in Table D.1.

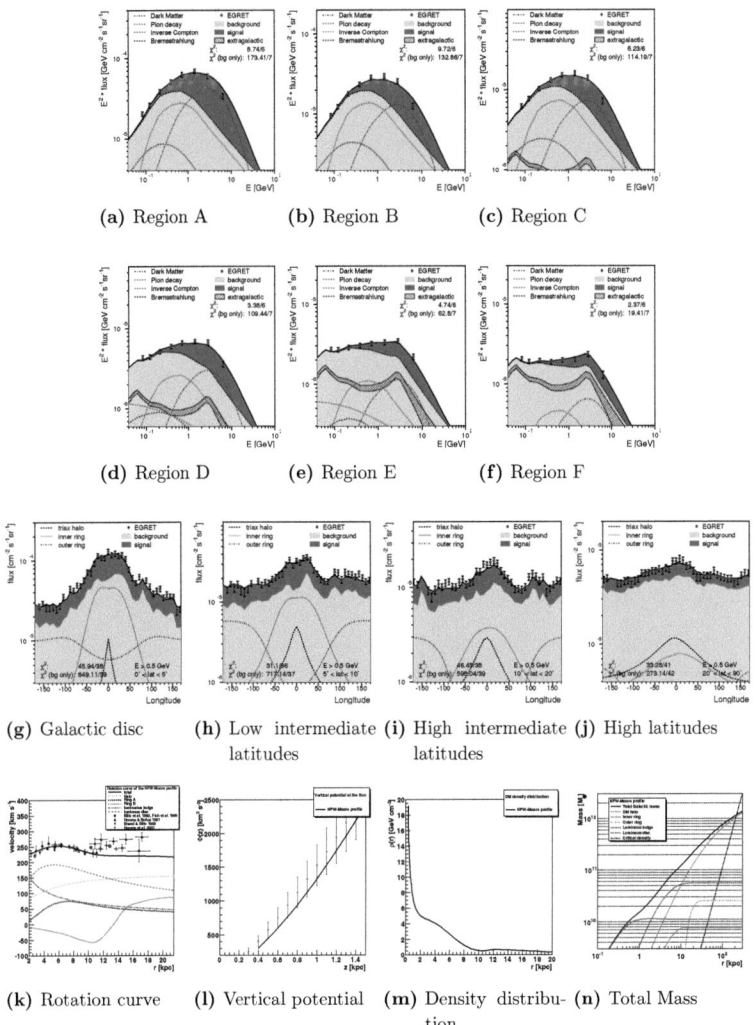

Figure D.3: Fit results for the NFW-Moore profile combination. The energy spectra of the diffuse Galactic gamma radiation are shown in **(a)** – **(f)** for the different regions defined in Section 4.2.1. The longitudinal distribution of the diffuse gamma radiation above a photon energy of 500 MeV is shown in **(g)** – **(j)**. The rotation curve in the Galactic disc, the gravitational potential perpendicular to the Galactic plane at the position of the Sun, the density distribution in the disc and the total mass of the Galaxy are shown in **(k)** – **(n)**. The parameters of the density distribution are given in Table D.1.

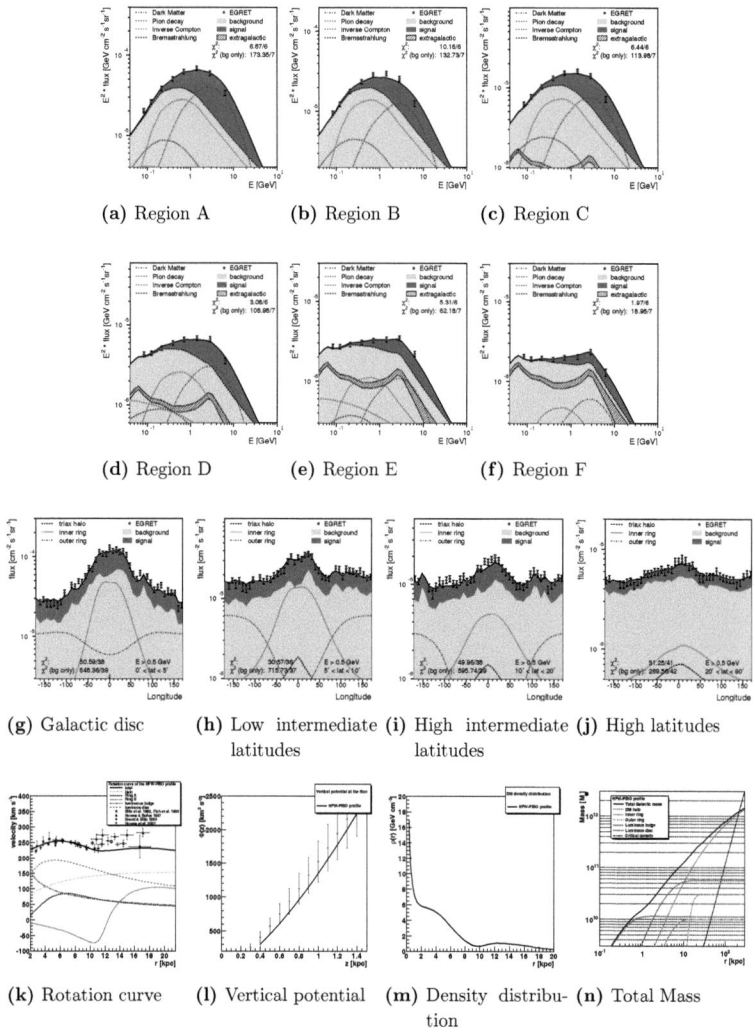

Figure D.4: Fit results for the NFW-PISO profile combination. The energy spectra of the diffuse Galactic gamma radiation are shown in (a) – (f) for the different regions defined in Section 4.2.1. The longitudinal distribution of the diffuse gamma radiation above a photon energy of 500 MeV is shown in (g) – (j). The rotation curve in the Galactic disc, the gravitational potential perpendicular to the Galactic plane at the position of the Sun, the density distribution in the disc and the total mass of the Galaxy are shown in (k) – (n). The parameters of the density distribution are given in Table D.1.

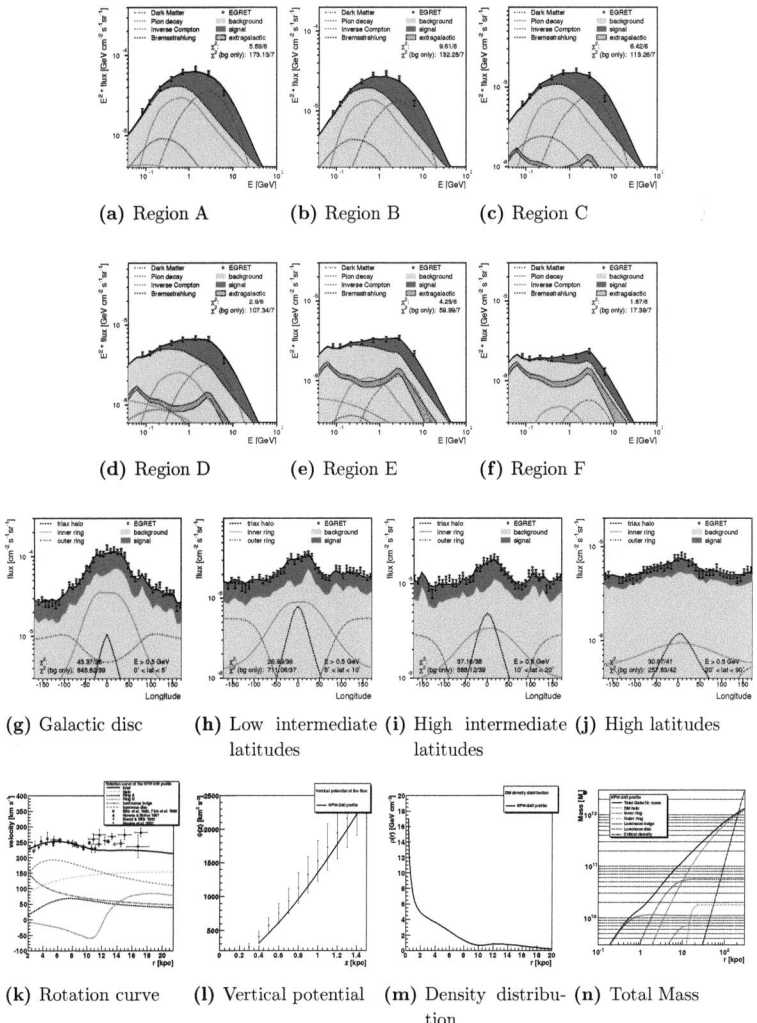

Figure D.5: Fit results for the NFW-240 profile combination. The energy spectra of the diffuse Galactic gamma radiation are shown in **(a)** – **(f)** for the different regions defined in Section 4.2.1. The longitudinal distribution of the diffuse gamma radiation above a photon energy of 500 MeV is shown in **(g)** – **(j)**. The rotation curve in the Galactic disc, the gravitational potential perpendicular to the Galactic plane at the position of the Sun, the density distribution in the disc and the total mass of the Galaxy are shown in **(k)** – **(n)**. The parameters of the density distribution are given in Table 4.4.

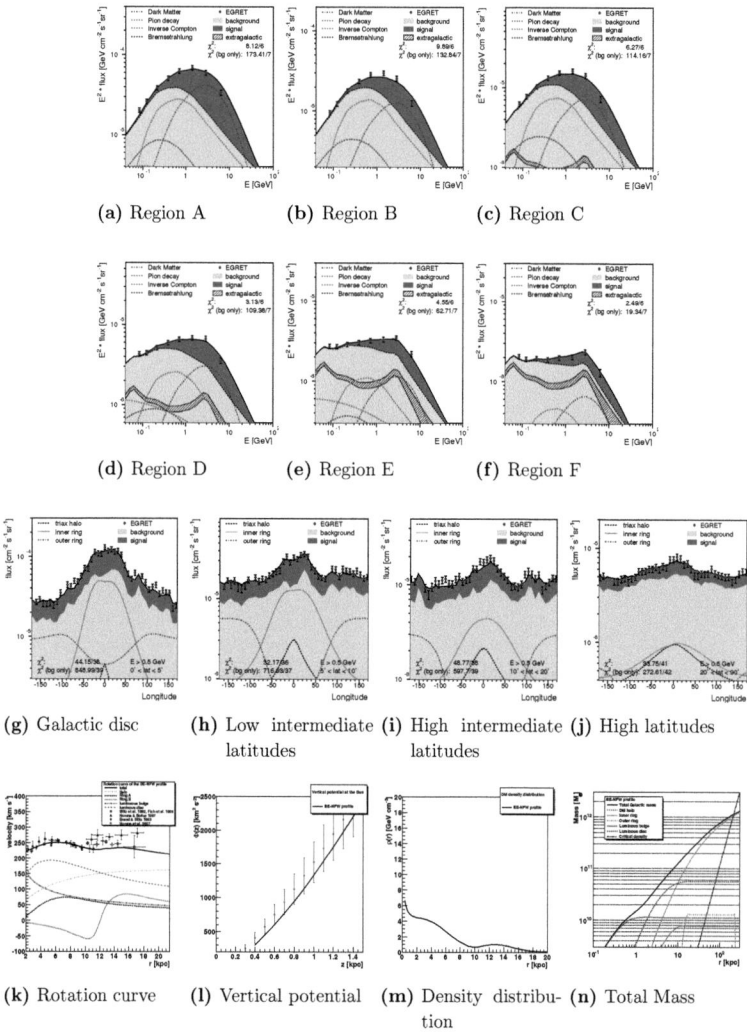

Figure D.6: Fit results for the BE-NFW profile combination. The energy spectra of the diffuse Galactic gamma radiation are shown in (a) – (f) for the different regions defined in Section 4.2.1. The longitudinal distribution of the diffuse gamma radiation above a photon energy of 500 MeV is shown in (g) – (j). The rotation curve in the Galactic disc, the gravitational potential perpendicular to the Galactic plane at the position of the Sun, the density distribution in the disc and the total mass of the Galaxy are shown in (k) – (n). The parameters of the density distribution are given in Table D.1.

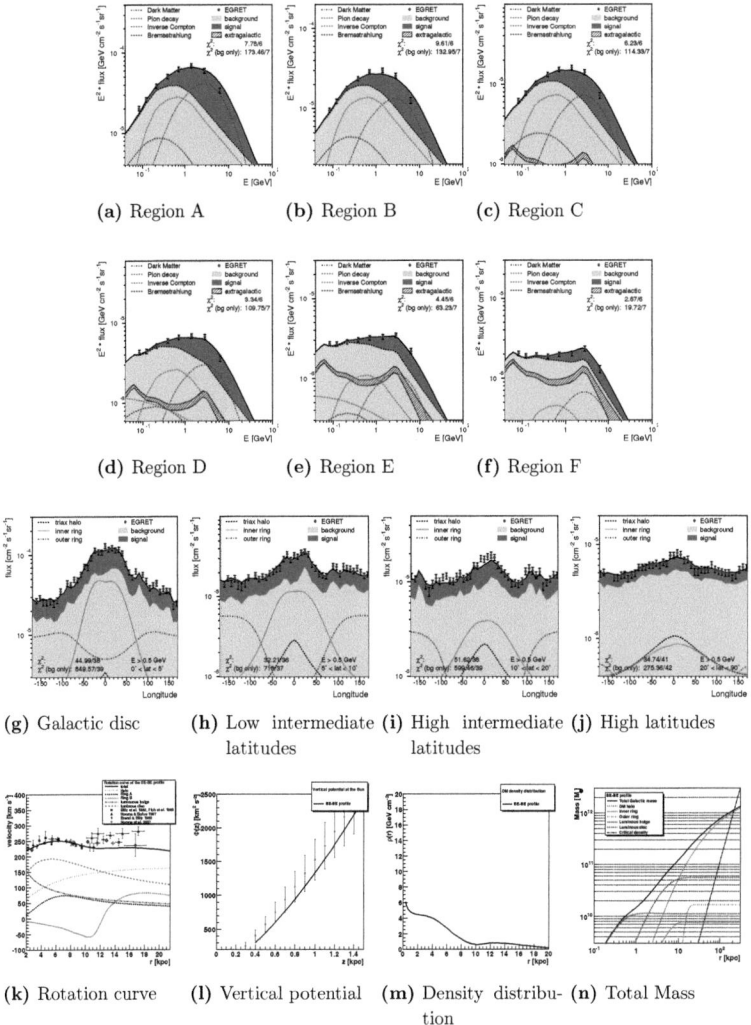

Figure D.7: Fit results for the BE-BE profile combination. The energy spectra of the diffuse Galactic gamma radiation are shown in **(a)** – **(f)** for the different regions defined in Section 4.2.1. The longitudinal distribution of the diffuse gamma radiation above a photon energy of 500 MeV is shown in **(g)** – **(j)**. The rotation curve in the Galactic disc, the gravitational potential perpendicular to the Galactic plane at the position of the Sun, the density distribution in the disc and the total mass of the Galaxy are shown in **(k)** – **(n)**. The parameters of the density distribution are given in Table D.1.

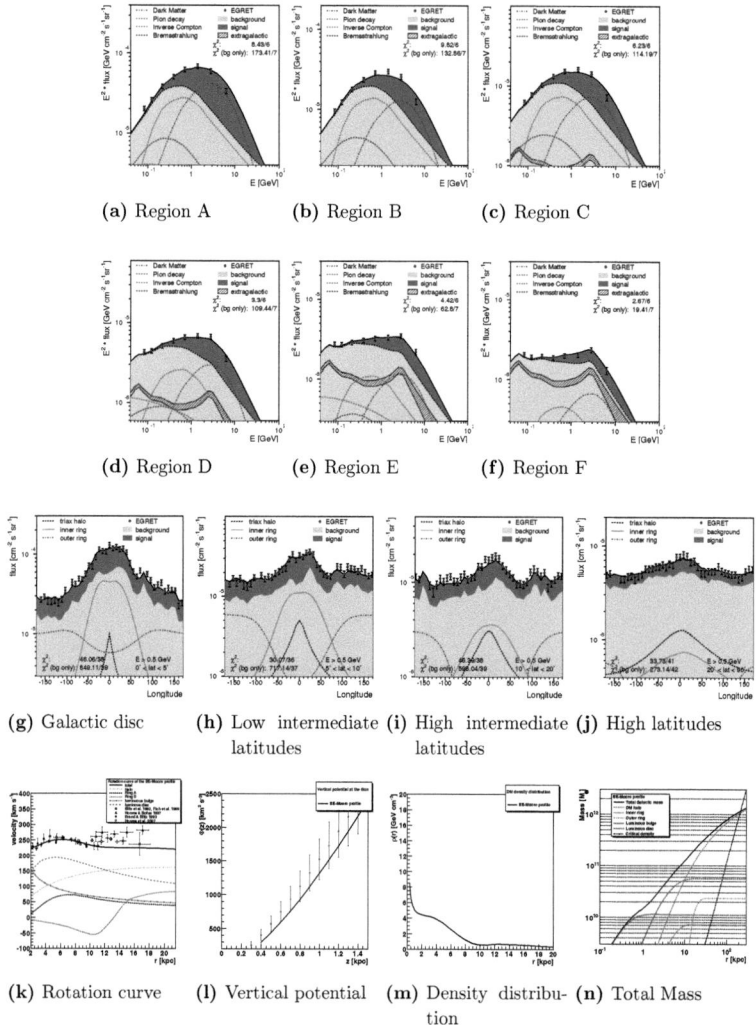

(a) Region A (b) Region B (c) Region C
(d) Region D (e) Region E (f) Region F
(g) Galactic disc (h) Low intermediate latitudes (i) High intermediate latitudes (j) High latitudes
(k) Rotation curve (l) Vertical potential (m) Density distribution (n) Total Mass

Figure D.8: Fit results for the BE-Moore profile combination. The energy spectra of the diffuse Galactic gamma radiation are shown in **(a)** – **(f)** for the different regions defined in Section 4.2.1. The longitudinal distribution of the diffuse gamma radiation above a photon energy of 500 MeV is shown in **(g)** – **(j)**. The rotation curve in the Galactic disc, the gravitational potential perpendicular to the Galactic plane at the position of the Sun, the density distribution in the disc and the total mass of the Galaxy are shown in **(k)** – **(n)**. The parameters of the density distribution are given in Table D.1.

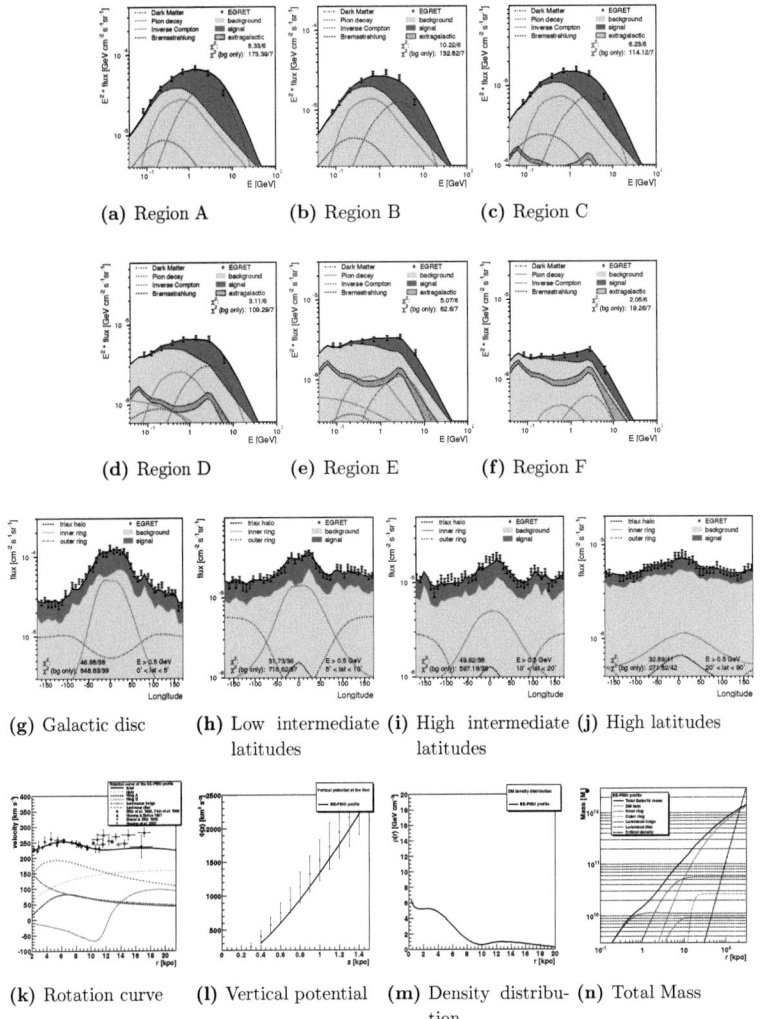

Figure D.9: Fit results for the BE-PISO profile combination. The energy spectra of the diffuse Galactic gamma radiation are shown in **(a)** – **(f)** for the different regions defined in Section 4.2.1. The longitudinal distribution of the diffuse gamma radiation above a photon energy of 500 MeV is shown in **(g)** – **(j)**. The rotation curve in the Galactic disc, the gravitational potential perpendicular to the Galactic plane at the position of the Sun, the density distribution in the disc and the total mass of the Galaxy are shown in **(k)** – **(n)**. The parameters of the density distribution are given in Table D.1.

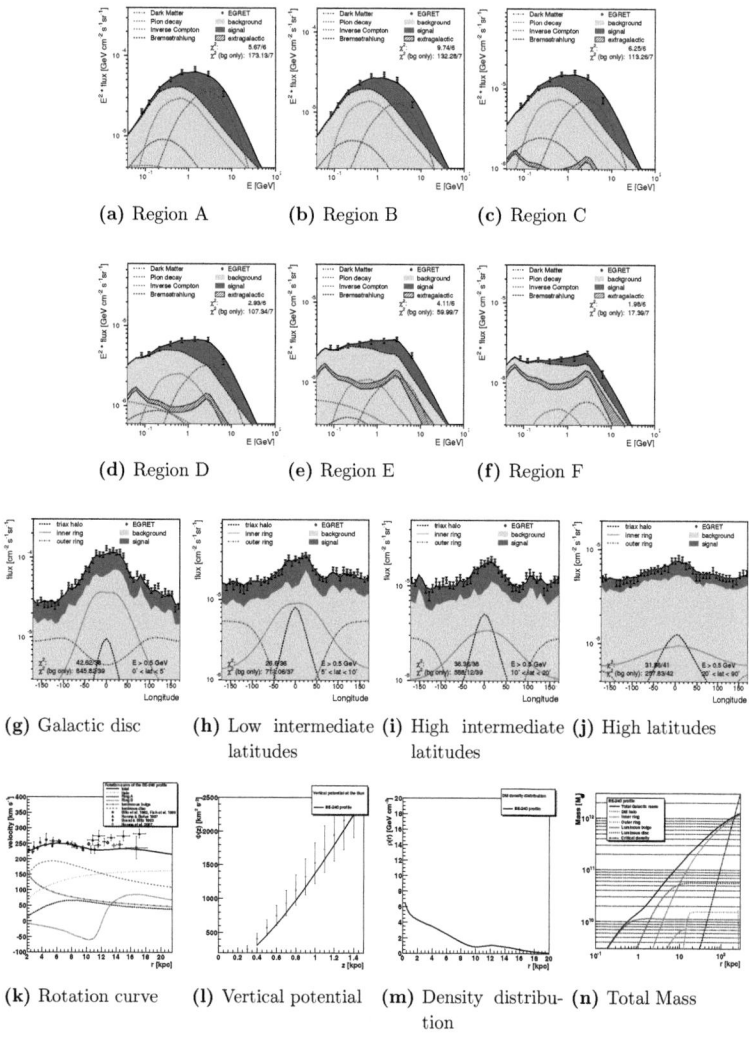

Figure D.10: Fit results for the BE-240 profile combination. The energy spectra of the diffuse Galactic gamma radiation are shown in (a) – (f) for the different regions defined in Section 4.2.1. The longitudinal distribution of the diffuse gamma radiation above a photon energy of 500 MeV is shown in (g) – (j). The rotation curve in the Galactic disc, the gravitational potential perpendicular to the Galactic plane at the position of the Sun, the density distribution in the disc and the total mass of the Galaxy are shown in (k) – (n). The parameters of the density distribution are given in Table 4.4.

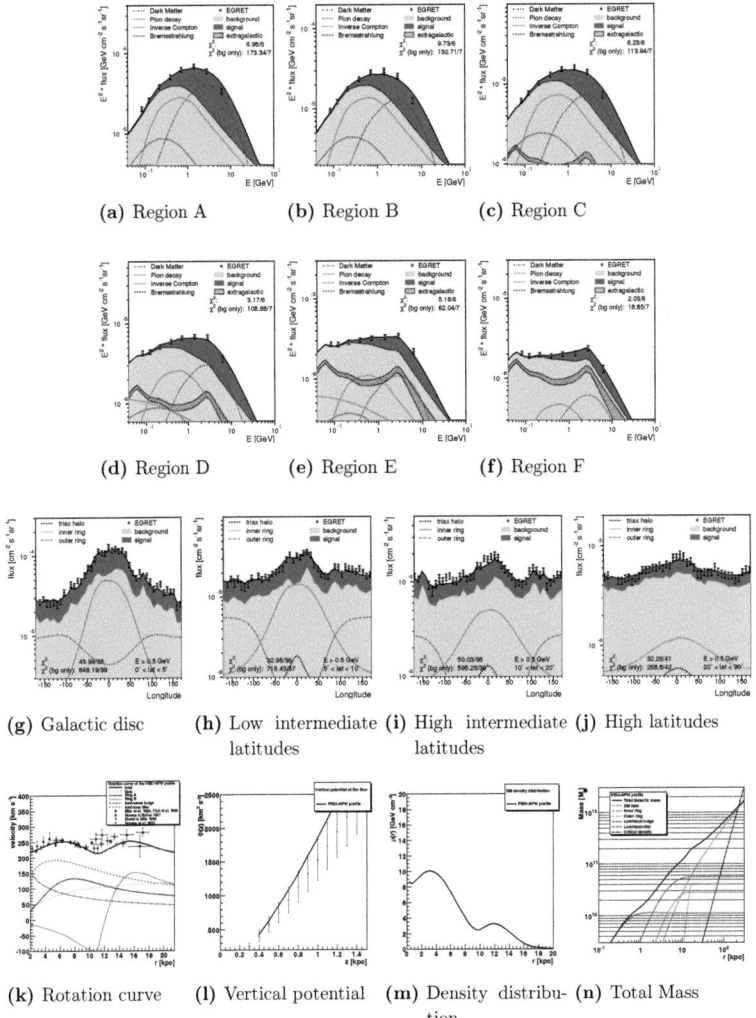

Figure D.11: Fit results for the PISO-NFW profile combination. The energy spectra of the diffuse Galactic gamma radiation are shown in (a) – (f) for the different regions defined in Section 4.2.1. The longitudinal distribution of the diffuse gamma radiation above a photon energy of 500 MeV is shown in (g) – (j). The rotation curve in the Galactic disc, the gravitational potential perpendicular to the Galactic plane at the position of the Sun, the density distribution in the disc and the total mass of the Galaxy are shown in (k) – (n). The parameters of the density distribution are given in Table D.1.

150　　　　　　　　　　　　　　　　Double Profile plots

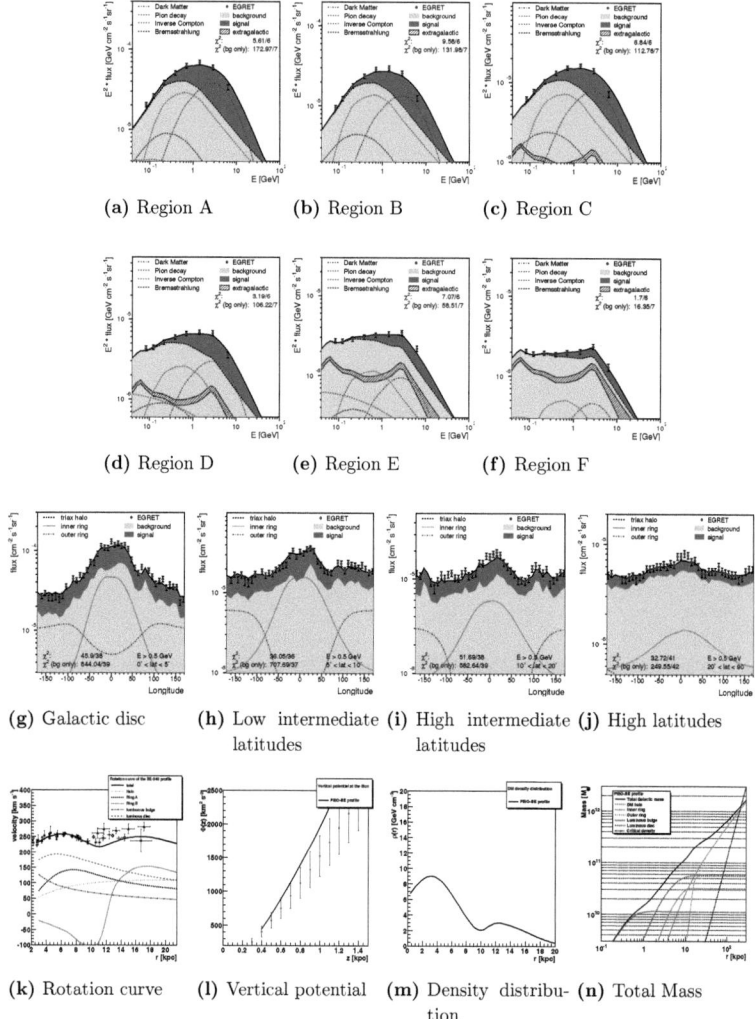

(a) Region A　(b) Region B　(c) Region C

(d) Region D　(e) Region E　(f) Region F

(g) Galactic disc　(h) Low intermediate latitudes　(i) High intermediate latitudes　(j) High latitudes

(k) Rotation curve　(l) Vertical potential　(m) Density distribution　(n) Total Mass

Figure D.12: Fit results for the PISO-BE profile combination. The energy spectra of the diffuse Galactic gamma radiation are shown in (a) – (f) for the different regions defined in Section 4.2.1. The longitudinal distribution of the diffuse gamma radiation above a photon energy of 500 MeV is shown in (g) – (j). The rotation curve in the Galactic disc, the gravitational potential perpendicular to the Galactic plane at the position of the Sun, the density distribution in the disc and the total mass of the Galaxy are shown in (k) – (n). The parameters of the density distribution are given in Table D.1.

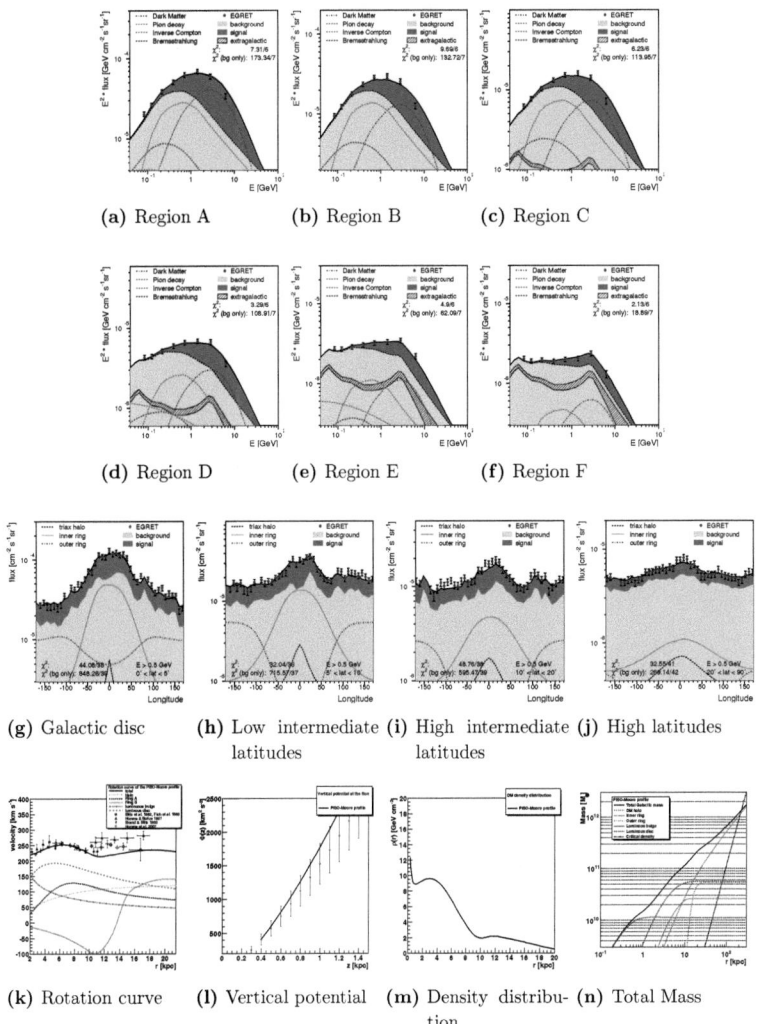

Figure D.13: Fit results for the PISO-Moore profile combination. The energy spectra of the diffuse Galactic gamma radiation are shown in **(a)** – **(f)** for the different regions defined in Section 4.2.1. The longitudinal distribution of the diffuse gamma radiation above a photon energy of 500 MeV is shown in **(g)** – **(j)**. The rotation curve in the Galactic disc, the gravitational potential perpendicular to the Galactic plane at the position of the Sun, the density distribution in the disc and the total mass of the Galaxy are shown in **(k)** – **(n)**. The parameters of the density distribution are given in Table D.1.

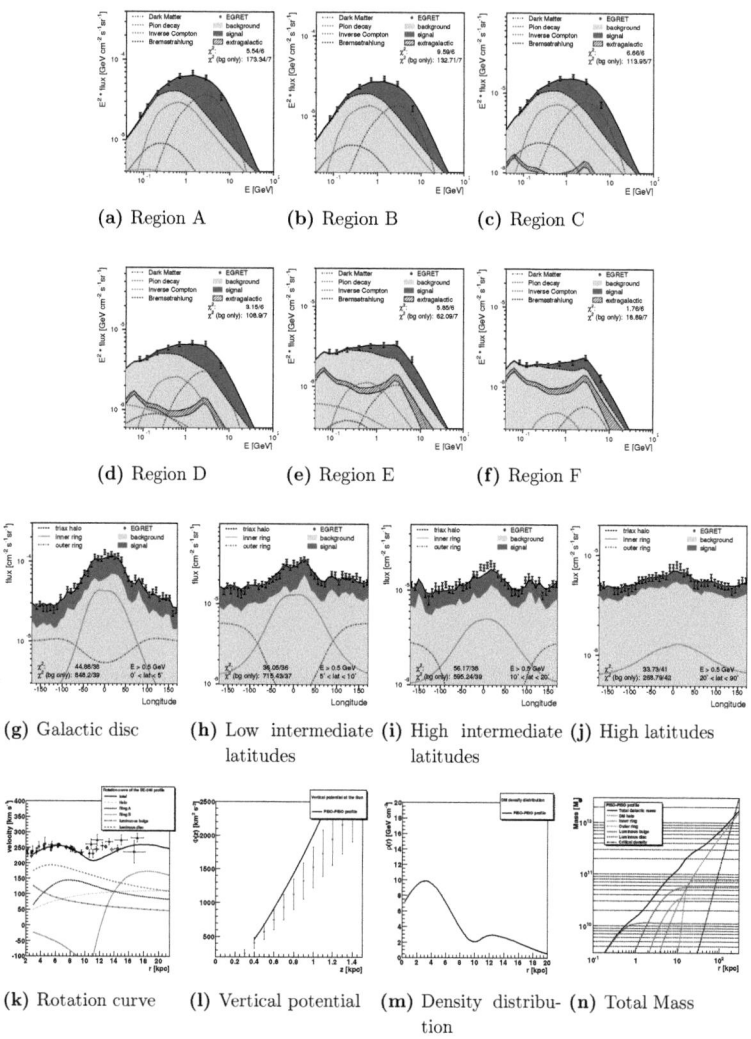

(a) Region A (b) Region B (c) Region C
(d) Region D (e) Region E (f) Region F
(g) Galactic disc (h) Low intermediate latitudes (i) High intermediate latitudes (j) High latitudes
(k) Rotation curve (l) Vertical potential (m) Density distribution (n) Total Mass

Figure D.14: Fit results for the PISO-PISO profile combination. The energy spectra of the diffuse Galactic gamma radiation are shown in **(a)** – **(f)** for the different regions defined in Section 4.2.1. The longitudinal distribution of the diffuse gamma radiation above a photon energy of 500 MeV is shown in **(g)** – **(j)**. The rotation curve in the Galactic disc, the gravitational potential perpendicular to the Galactic plane at the position of the Sun, the density distribution in the disc and the total mass of the Galaxy are shown in **(k)** – **(n)**. The parameters of the density distribution are given in Table D.1.

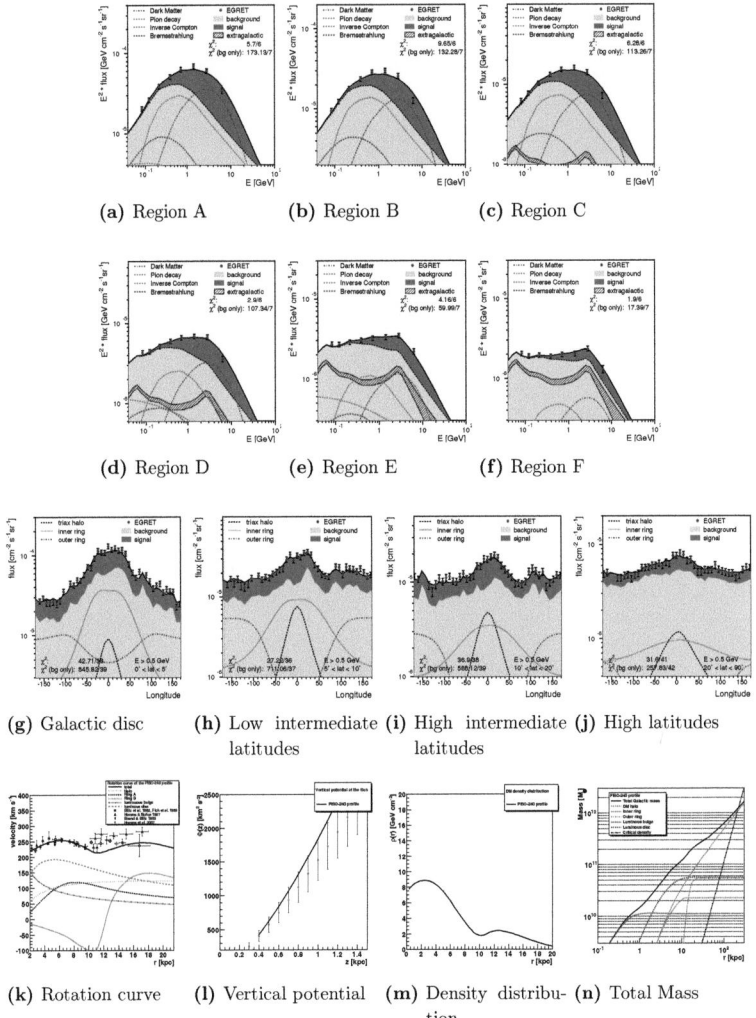

Figure D.15: Fit results for the PISO-240 profile combination. The energy spectra of the diffuse Galactic gamma radiation are shown in **(a)** – **(f)** for the different regions defined in Section 4.2.1. The longitudinal distribution of the diffuse gamma radiation above a photon energy of 500 MeV is shown in **(g)** – **(j)**. The rotation curve in the Galactic disc, the gravitational potential perpendicular to the Galactic plane at the position of the Sun, the density distribution in the disc and the total mass of the Galaxy are shown in **(k)** – **(n)**. The parameters of the density distribution are given in Table 4.4.

Bibliography

[1] F. Zwicky, *Spectral displacement of extra galactic nebulae*, Helv. Phys. Acta **6** (1933) 110–127. in german. 1, 34

[2] Y. Sofue and V. Rubin, *Rotation Curves of Spiral Galaxies*, Ann. Rev. Astron. Astrophys. **39** (2001) 137–174, [`astro-ph/0010594`]. 1

[3] **WMAP** Collaboration, E. Komatsu et. al., *Five-Year Wilkinson Microwave Anisotropy Probe (WMAP) Observations:Cosmological Interpretation*, Astrophys. J. Suppl. **180** (2009) 330–376, [`0803.0547`]. 1, 5, 13, 14, 16, 19, 36

[4] G. Steigman, *Cosmology Confronts Particle Physics*, Ann. Rev. Nucl. Part. Sci. **29** (1979) 313–338. 2

[5] G. Jungman, M. Kamionkowski, and K. Griest, *Supersymmetric dark matter*, Phys. Rept. **267** (1996) 195–373, [`hep-ph/9506380`]. 2

[6] E. W. Kolb and M. S. Turner, *The early Universe*. Westview Press, 1990. 2, 12, 22, 23, 25, 29, 40

[7] J. Rich, *Fundamentals of Cosmology*. Springer-Verlag Berlin Heidelberg, 2001. 2, 5, 22, 25, 29

[8] D.-E. Liebscher, *Cosmology*. Springer Verlag, Berlin, Germany, 2005. 2

[9] V. Berezinsky, V. Dokuchaev, and Y. Eroshenko, *Cosmological Origin of Small-Scale Clumps and DM Annihilation Signal*, `astro-ph/0412305`. 2

[10] V. Berezinsky, V. Dokuchaev, and Y. Eroshenko, *Remnants of dark matter clumps*, Phys. Rev. **D77** (2008) 083519, [`0712.3499`]. 2, 41, 42, 44, 80

[11] V. Springel et. al., *The Aquarius Project: the subhalos of galactic halos*, `0809.0898`. 2, 42

[12] M. Kuhlen, J. Diemand, P. Madau, and M. Zemp, *The Via Lactea INCITE Simulation: Galactic Dark Matter Substructure at High Resolution*, J. Phys. Conf. Ser. **125** (2008) 012008, [0810.3614]. 2, 41, 42

[13] A. D. Ludlow et. al., *The Unorthodox Orbits of Substructure Halos*, 0801.1127. 2, 3, 82

[14] N. J. Spooner, *Direct Dark Matter Searches*, J. Phys. Soc. Jap. **76** (2007) 111016, [0705.3345]. 2, 62

[15] W. de Boer, *Indirect Dark Matter Searches in the Light of ATIC, FERMI, EGRET and PAMELA*, 0910.2601. 2

[16] W. de Boer, C. Sander, V. Zhukov, A. V. Gladyshev, and D. I. Kazakov, *EGRET excess of diffuse galactic gamma rays as tracer of dark matter*, Astron. Astrophys. **444** (2005) 51, [astro-ph/0508617]. 2, 45, 55, 62, 78, 80, 81, 84, 85, 88, 90, 111

[17] L. Bergstrom, J. Edsjo, M. Gustafsson, and P. Salati, *Is the dark matter interpretation of the EGRET gamma excess compatible with antiproton measurements?*, JCAP **0605** (2006) 006, [astro-ph/0602632]. 2, 99, 112

[18] **SDSS** Collaboration, H. J. Newberg et. al., *The Ghost of Sagittarius and Lumps in the Halo of the Milky Way*, Astrophys. J. **569** (2002) 245–274, [astro-ph/0111095]. 3, 31

[19] R. A. Ibata, M. J. Irwin, G. F. Lewis, A. M. N. Ferguson, and N. Tanvir, *One Ring to Encompass them All: A giant stellar structure that surrounds the Galaxy*, Mon. Not. Roy. Astron. Soc. **340** (2003) L21, [astro-ph/0301067]. 3, 31

[20] P. M. W. Kalberla, L. Dedes, J. Kerp, and U. Haud, *Dark matter in the Milky Way, II. the HI gas distribution as a tracer of the gravitational potential*, 0704.3925. 3, 53, 59, 71

[21] J. A. Peacock et. al., *A measurement of the cosmological mass density from clustering in the 2dF Galaxy Redshift Survey*, Nature **410** (2001) 169–173, [astro-ph/0103143]. 6

[22] M. Colless et. al., *The 2dF Galaxy Redshift Survey: Final Data Release*, astro-ph/0306581. 6

[23] A. Einstein, *The foundation of the general theory of relativity*, Annalen Phys. **49** (1916) 769–822. 7

[24] A. Einstein, *Cosmological Considerations in the General Theory of Relativity*, Sitzungsber. Preuss. Akad. Wiss. Berlin (Math. Phys.) **1917** (1917) 142–152. 8

[25] A. A. Friedmann, *ber die Krmmung des Raumes*, Zeitschrift fr Physik **10** (1922) 377–386. 8

[26] A. Einstein, *On the electrodynamics of moving bodies*, Annalen Phys. **17** (1905) 891–921. 8

[27] NASA — Astronomy and Cosmology
http://rst.gsfc.nasa.gov/Sect20/A1.html
Date: January 18th, 2010. 11

[28] H. S. Kragh, *Conceptions of cosmos: from myths to the accelerating universe; a history of cosmology*. Oxford University Press, 2007. 14

[29] G. F. Smoot et. al., *Structure in the COBE differential microwave radiometer first year maps*, Astrophys. J. **396** (1992) L1–L5. 14

[30] **WMAP** Collaboration, G. Hinshaw et. al., *Five-Year Wilkinson Microwave Anisotropy Probe (WMAP) Observations:Data Processing, Sky Maps, & Basic Results*, Astrophys. J. Suppl. **180** (2009) 225–245, [0803.0732]. 15

[31] **WMAP** Collaboration, M. R. Nolta et. al., *Five-Year Wilkinson Microwave Anisotropy Probe (WMAP) Observations: Angular Power Spectra*, Astrophys. J. Suppl. **180** (2009) 296–305, [0803.0593]. 16

[32] **Supernova Cosmology Project** Collaboration, M. Kowalski et. al., *Improved Cosmological Constraints from New, Old and Combined Supernova Datasets*, Astrophys. J. **686** (2008) 749–778, [0804.4142]. 17, 18

[33] W. J. Percival et. al., *Measuring the Baryon Acoustic Oscillation scale using the SDSS and 2dFGRS*, Mon. Not. Roy. Astron. Soc. **381** (2007) 1053–1066, [0705.3323]. 17

[34] G. Gamow, *Expanding universe and the origin of elements*, Phys. Rev. **70** (1946) 572–573. 20

[35] F. Hoyle and R. J. Tayler, *The Mystery of the Cosmic Helium Abundance*, Nature **203** (1964) 1108. 20

[36] R. V. Wagoner, *Big bang nucleosynthesis revisited*, Astrophys. J. **179** (1973) 343–360. 20

[37] A. H. Guth and E. J. Weinberg, *Cosmological Consequences of a First Order Phase Transition in the SU(5) Grand Unified Model*, Phys. Rev. **D23** (1981) 876. 23

[38] A. D. Linde, *A New Inflationary Universe Scenario: A Possible Solution of the Horizon, Flatness, Homogeneity, Isotropy and Primordial Monopole Problems*, Phys. Lett. **B108** (1982) 389–393. 23

[39] A. J. Albrecht and P. J. Steinhardt, *Cosmology for Grand Unified Theories with Radiatively Induced Symmetry Breaking*, Phys. Rev. Lett. **48** (1982) 1220–1223. 23

[40] J. Diemand et. al., *Clumps and streams in the local dark matter distribution*, Nature **454** (2008) 735–738, [0805.1244]. 27, 41

[41] Y. Zhang et. al., *The spin and orientation of dark matter halos within cosmic filaments*, Astrophys. J. **706** (2009) 747–761, [0906.1654]. 27

[42] Max-Planck Intitute of Astrophysics — Galaxy – Dark Matter connection
http://www.mpa-garching.mpg.de/HIGHLIGHT/2002/highlight0206_e.html
Date: January 18th, 2010. 28

[43] S. Weinberg, *Gravitation and Cosmology - Principles and applications of the general theory of relativity*. John Wiley & Sons, Inc., 1972. 29

[44] E. P. Hubble, *The Realm of the Nebulae*. New Haven: Yale University Press, 1936. 29

[45] Atlas of the Universe
http://www.atlasoftheuniverse.com
Date: January 18th, 2010. 30

[46] R. A. Ibata, G. Gilmore, and M. J. Irwin, *A Dwarf satellite galaxy in Sagittarius*, Nature **370** (1994) 194. 30

[47] N. F. Martin et. al., *A dwarf galaxy remnant in Canis Major: the fossil of an in-plane accretion onto the Milky Way*, Mon. Not. Roy. Astron. Soc. **348** (2004) 12, [astro-ph/0311010]. 30, 31, 69

[48] J. S. Sparke, L. S. & Gallagher, *Galaxies in the Universe - An Introduction*. Cambridge University Press, 2007. 31, 34, 51, 60

[49] J. H. Oort, *The force exerted by the stellar system in the direction perpendicular to the galactic plane and some related problems*, Bulletin of the Astronomical Institutes of the Netherlands **6** (1932) 249. 34

[50] J. Ostriker, P. Peebles, and A. Yahil, *The Size and Mass of Galaxies, and the Mass of the Universe*, Astrophysical Journal **193** (1974) L1–L4. 34

[51] J. Einasto, A. Kaasik, and E. Saar, *Dynamical evidence for massive coronas of galaxies*, Nature **250** (1974) 309–310. 34

[52] M. Taoso, G. Bertone, and A. Masiero, *Dark Matter Candidates: A Ten-Point Test*, JCAP **0803** (2008) 022, [0711.4996]. 35

[53] G. Bertone, D. Hooper, and J. Silk, *Particle dark matter: Evidence, candidates and constraints*, Phys. Rept. **405** (2005) 279–390, [hep-ph/0404175]. 35

[54] L. Bergstrom, *Non-baryonic dark matter: Observational evidence and detection methods*, Rept. Prog. Phys. **63** (2000) 793, [hep-ph/0002126]. 35, 36

[55] J. Bekenstein and M. Milgrom, *Does the missing mass problem signal the breakdown of Newtonian gravity?*, Astrophys. J. **286** (1984) 7–14. 35

[56] O. Bienayme, B. Famaey, X. Wu, H. S. Zhao, and D. Aubert, *Galactic kinematics with modified Newtonian dynamics*, arXiv:astro-ph/0904.3893 (2009) [0904.3893]. 35

[57] B. Paczynski, *Gravitational microlensing by the galactic halo*, Astrophys. J. **304** (1986) 1–5. 35

[58] E. Aubourg et. al., *Evidence for gravitational microlensing by dark objects in the galactic halo*, Nature **365** (1993) 623–625. 35

[59] Y. Suzuki, *Kamiokande solar neutrino results*, Nucl. Phys. Proc. Suppl. **38** (1995) 54–59. 36

[60] **SNO** Collaboration, Q. R. Ahmad et. al., *Direct evidence for neutrino flavor transformation from neutral-current interactions in the Sudbury Neutrino Observatory*, Phys. Rev. Lett. **89** (2002) 011301, [nucl-ex/0204008]. 36

[61] **K2K** Collaboration, M. H. Ahn et. al., *Measurement of Neutrino Oscillation by the K2K Experiment*, Phys. Rev. **D74** (2006) 072003, [hep-ex/0606032]. 36

[62] G. Drexlin, *Final neutrino oscillation results from LSND and KARMEN*, Nucl. Phys. Proc. Suppl. **118** (2003) 146–153. 36

[63] J. N. Bahcall, M. H. Pinsonneault, S. Basu, and J. Christensen-Dalsgaard, *Are standard solar models reliable?*, Phys. Rev. Lett. **78** (1997) 171–174, [astro-ph/9610250]. 36

[64] C. Weinheimer, *The neutrino mass direct measurements*, in *Proceedings of 10th International Workshop on Neutrino Telescopes*, 2003. hep-ex/0306057. 36

[65] **KATRIN** Collaboration, A. Osipowicz et. al., *KATRIN: A next generation tritium beta decay experiment with sub-eV sensitivity for the electron neutrino mass*, hep-ex/0109033. 36

[66] L. Bergstrom, *Dark Matter Candidates*, New J. Phys. **11** (2009) 105006, [0903.4849]. 36

[67] R. D. Peccei and H. R. Quinn, *CP Conservation in the Presence of Instantons*, Phys. Rev. Lett. **38** (1977) 1440–1443. 36

[68] D. I. Kazakov, *Beyond the standard model (in search of supersymmetry)*, hep-ph/0012288. 37

[69] K. Griest and M. Kamionkowski, *Unitarity Limits on the Mass and Radius of Dark Matter Particles*, Phys. Rev. Lett. **64** (1990) 615. 37

[70] E. W. Kolb, D. J. H. Chung, and A. Riotto, *WIMPzillas!*, hep-ph/9810361. 38

[71] T. Kaluza, *Zum Unitarittsproblem der Physik*, Sitzungsberichte Preuische Akademie der Wissenschaften **96** (1921) 69. 38

[72] M. Regis, M. Serone, and P. Ullio, *A dark matter candidate from an extra (non-universal) dimension*, JHEP **03** (2007) 084, [hep-ph/0612286]. 38

[73] P. Gondolo et. al., *DarkSUSY: Computing supersymmetric dark matter properties numerically*, JCAP **0407** (2004) 008, [astro-ph/0406204]. 39

[74] G. Belanger, F. Boudjema, A. Pukhov, and A. Semenov, *micrOMEGAs 2.0.7: A program to calculate the relic density of dark matter in a generic model*, Comput. Phys. Commun. **177** (2007) 894–895. 39

[75] C. Sander, *Interpretation of the Excess in Diffuse Galactic Gamma Rays above 1 GeV as Dark Matter Annihilation Signal*. PhD thesis, University of Karlsruhe, Germany, 2005. 40, 45, 76

[76] B. Moore et. al., *Dark matter substructure within galactic halos*, Astrophys. J. **524** (1999) L19–L22. 41, 61, 62

[77] T. Sjostrand, S. Mrenna, and P. Skands, *PYTHIA 6.4 Physics and Manual*, JHEP **05** (2006) 026, [hep-ph/0603175]. 45

[78] A. W. Strong, I. V. Moskalenko, and O. Reimer, *Diffuse Galactic continuum gamma rays. A model compatible with EGRET data and cosmic-ray measurements*, Astrophys. J. **613** (2004) 962–976, [astro-ph/0406254]. 45

[79] A. W. Strong and I. V. Moskalenko, *The GALPROP program for cosmic-ray propagation: new developments*, astro-ph/9906228. 46, 78

[80] V. N. Zirakashvili et. al., *Magnetohydrodynamic wind driven by cosmic rays in a rotating galaxy*, Astronomy and Astrophysics **311** (1996) 113. 46

[81] W. de Boer, A. Nordt, C. Sander, and V. Zhukov, *A new Determination of the Extragalactic Background of Diffuse Gamma Rays taking into account Dark Matter Annihilation*, 0705.0094. 47, 81

[82] M. R. Binney, E. & Merrifield, *Galactic astronomy*. Princeton University Press, 1998. 51

[83] S. Zeilik, M. & Gregory, *Introductory Astronomy and Astrophysics*. Saunders College, 1998. 51

[84] F. J. Kerr and D. Lynden-Bell, *Review of galactic constants*, Mon. Not. Roy. Astron. Soc. **221** (1986) 1023. 51

[85] M. J. Reid and A. Brunthaler, *The Proper Motion of Sgr A*: II. The Mass of Sgr A**, Astrophys. J. **616** (2004) 872–884, [astro-ph/0408107]. 51

[86] M. J. Reid et. al., *Trigonometric Parallaxes of Massive Star Forming Regions: VI. Galactic Structure, Fundamental Parameters and Non- Circular Motions*, Astrophys. J. **700** (2009) 137–148, [0902.3913]. 52

[87] B. Fuchs et. al., *The kinematics of late type stars in the solar cylinder studied with SDSS data*, Astron. J. **137** (2009) 4149–4159, [0902.2324]. 52

[88] R. P. Olling and W. Dehnen, *The Oort Constants Measured from Proper Motions*, Astrophys. J. **599** (2003) 275–296, [astro-ph/0301486]. 52

[89] S. Gillessen et. al., *Monitoring stellar orbits around the Massive Black Hole in the Galactic Center*, Astrophys. J. **692** (2009) 1075–1109, [0810.4674]. 52

[90] A. M. Ghez et. al., *Measuring Distance and Properties of the Milky Way's Central Supermassive Black Hole with Stellar Orbits*, arXiv:astro-ph/0808.2870 (2008) [0808.2870]. 52

[91] M. R. Merrifield, *The Rotation curve of the milky way to 2.5-R(0) from the thickness of the HI layer*, Astron. J. **103** (1992). CITA-91-44. 53, 55

[92] Y. Sofue, M. Honma, and T. Omodaka, *Unified Rotation Curve of the Galaxy – Decomposition into de Vaucouleurs Bulge, Disk, Dark Halo, and the 9-kpc Rotation Dip –*, 0811.0859. 54, 55, 69

[93] Burton, W. P. and Gordon, M. A., *Carbon monoxide in the Galaxy. III - The overall nature of its distribution in the equatorial plane*, Astron. Astrophys. **63** (1978) 7–27. 54, 119, 121

[94] Clemens, D. P., *Massachusetts-Stony Brook Galactic plane CO survey - The Galactic disk rotation curve*, Astrophys. J. **295** (1985) 422–428. 54, 104, 119, 122

[95] Fich, Michel and Blitz, Leo and Stark, Antony A., *The rotation curve of the Milky Way to 2 R(0)*, Astrophys. J. **342** (1989) 272–284. 54, 119, 124

[96] Blitz, Leo and Fich, Michel and Stark, Antony A., *Catalog of CO radial velocities toward galactic H II regions*, Astrophys. J. **49** (1982) 183–206. 54, 119, 125

[97] Honma, M. and Sofue, Y., *Rotation Curve of the Galaxy*, PASJ **49** (1997) 453–460. 54

[98] Brand, J. and Blitz, L., *The Velocity Field of the Outer Galaxy*, Astron. Astrophys. **275** (1993) 67. 54, 55, 119, 126

[99] Demers, S. and Battinelli, P., *C stars as kinematic probes of the Milky Way disk from 9 to 15 kpc*, Astron. Astrophys. **473** (2007) 143–148. 55, 119, 123

[100] **SDSS** Collaboration, X. X. Xue et. al., *The Milky Way's Circular Velocity Curve to 60 kpc and an Estimate of the Dark Matter Halo Mass from Kinematics of 2400 SDSS Blue Horizontal Branch Stars*, Astrophys. J. **684** (2008) 1143–1158, [0801.1232]. 56, 68

[101] M. I. Wilkinson and N. W. Evans, *The Present and Future Mass of the Milky Way Halo*, Mon. Not. Roy. Astron. Soc. **310** (1999) 645, [astro-ph/9906197]. 56

[102] T. Sakamoto, M. Chiba, and T. C. Beers, *The Mass of the Milky Way: Limits from a Newly Assembled Set of Halo Objects*, Astron. Astrophys. **397** (2003) 899–912, [astro-ph/0210508]. 56

[103] G. R. Blumenthal, S. M. Faber, R. Flores, and J. R. Primack, *Contraction of Dark Matter Galactic Halos Due to Baryonic Infall*, Astrophys. J. **301** (1986) 27. 56

[104] G. Battaglia et. al., *The radial velocity dispersion profile of the Galactic halo: Constraining the density profile of the dark halo of the Milky Way*, Mon. Not. Roy. Astron. Soc. **364** (2005) 433–442, [astro-ph/0506102]. 57

[105] A. Klypin, H. Zhao, and R. S. Somerville, *LCDM-based models for the Milky Way and M31 I: Dynamical Models*, Astrophys. J. **573** (2002) 597–613, [astro-ph/0110390]. 57

[106] T. Naab and J. P. Ostriker, *A simple model for the evolution of disc galaxies: The Milky Way*, Mon. Not. Roy. Astron. Soc. **366** (2006) 899–917, [astro-ph/0505594]. 57

[107] K. Kuijken and G. Gilmore, *The galactic disk surface mass density and the Galactic force K(z) at Z = 1.1 kiloparsecs*, Astrophys. J. **367** (1991) L9–L13. 57, 58

[108] J. Holmberg and C. Flynn, *The local surface density of disc matter mapped by Hipparcos*, Mon. Not. Roy. Astron. Soc. **352** (2004) 440, [astro-ph/0405155]. 57, 58, 59

[109] G. Gilmore, R. F. G. Wyse, and K. Kuijken, *Stellar dynamics and Galactic evolution*, in *Evolutionary phenomena in galaxies*, pp. 172–200, Cambridge University Press, 1989. 57

[110] A. Gould, J. N. Bahcall, and C. Flynn, *Disk M dwarf luminosity function from HST star counts*, Astrophys. J. **465** (1996) 759, [astro-ph/9505087]. 57

[111] Z. Zheng, C. Flynn, A. Gould, J. N. Bahcall, and S. Salim, *M Dwarfs from Hubble Space Telescope Star Counts. IV*, Astrophys. J. **555** (2001) 393–404, [astro-ph/0102442]. 57

[112] D. Mera, G. Chabrier, and R. Schaeffer, *Towards a consistent model of the Galaxy: I. kinematic properties, star counts and microlensing observations*, arXiv:astro-ph/9801051 (1998) [astro-ph/9801051]. 57

[113] T. M. Dame, *The Distribution of Neutral Gas in the Milky Way*, in *Back to the Galaxy*, AIP Conf. *278* (S. S. Holt and F. Verter, eds.), pp. 267–278, 1993. 57

[114] R. P. Olling and M. R. Merrifield, *Luminous and Dark Matter in the Milky Way*, Mon. Not. Roy. Astron. Soc. **326** (2001) 164, [astro-ph/0104465]. 57

[115] O. Bienayme, C. Soubiran, T. V. Mishenina, V. V. Kovtyukh, and A. Siebert, *Vertical distribution of Galactic disk stars: III. The Galactic disk surface mass density from red clump giants*, astro-ph/0510431. 58

[116] A. Siebert, O. Bienayme, and C. Soubiran, *Vertical distribution of Galactic disk stars : II. The surface mass density in the Galactic plane*, Astron. Astrophys. **399** (2003) 531–542, [astro-ph/0211328]. 58, 59

[117] V. I. Korchagin, T. M. Girard, T. V. Borkova, D. I. Dinescu, and W. F. van Altena, *Local Surface Density of the Galactic Disk from a 3-D Stellar Velocity Sample*, arXiv:astro-ph/0308276 (2003) [astro-ph/0308276]. 59

[118] G. S. Shostak and P. C. van der Kruit, *Study of nearly face-on spiral galaxies. II. HI synthesis observations and optical surface photometry of NGC 628*, Astron. Astrophys. **132** (1984) 20–32. 60

[119] B. M. Lewis, *Face-on galaxies*, Astrophys. J. **285** (1984) 453–457. 60

[120] V. F. Cardone and M. Sereno, *Modelling the Milky Way through adiabatic compression of cold dark matter halo*, arXiv:astro-ph/0501567 (2005) [astro-ph/0501567]. 60

[121] M. J. Reid, *Is there a Supermassive Black Hole at the Center of the Milky Way?*, Int. J. Mod. Phys. **D18** (2009) 889–910, [0808.2624]. 60

[122] F. Hammer, M. Puech, L. Chemin, H. Flores, and M. Lehnert, *The Milky Way: An Exceptionally Quiet Galaxy: Implications for the formation of spiral galaxies*, Astrophys. J. **662** (2007) 322–334, [astro-ph/0702585]. 60

[123] G. Gilmore and N. Reid, *New light on faint stars. III - Galactic structure towards the South Pole and the Galactic thick disc*, Mon. Not. Roy. Astron. Soc. **202** (1983) 1025–1047. 60

[124] P. Kroupa, C. A. Tout, and G. Gilmore, *The Distribution of low mass stars in the galactic disc*, Mon. Not. Roy. Astron. Soc. **262** (1993) 545. 61

[125] A. C. Robin, M. Haywood, M. Creze, D. K. Ojha, and O. Bienayme, *The thick disc of the galaxy: Sequel of a merging event*, Astron. Astrophys. **305** (1996) 125, [astro-ph/9504090]. 61

[126] D. Ojha, O. Bienayme, A. Robin, M. Creze, and V. Mohan, *Structure and kinematical properties of the Galaxy at intermediate galactic latitudes*, Astron. Astrophys. **311** (1996) 456–469, [astro-ph/9511049]. 61

[127] H. T. Freudenreich, *COBE's Galactic Bar and Disk*, Astrophys. J. **492** (1998) 495–510, [astro-ph/9707340]. 61

[128] Y. Ascasibar, P. Jean, C. Boehm, and J. Knoedlseder, *Constraints on dark matter and the shape of the Milky Way dark halo from the 511 keV line*, Mon. Not. Roy. Astron. Soc. **368** (2006) 1695–1705, [astro-ph/0507142]. 61, 62

[129] J. E. Gunn, *Massive galactic halos. I - Formation and evolution*, Astrophys. J. **218** (1977) 592. 61, 62

[130] J. F. Navarro, C. S. Frenk, and S. D. M. White, *The Structure of Cold Dark Matter Halos*, Astrophys. J. **462** (1996) 563–575, [astro-ph/9508025]. 61, 62

[131] M. Ricotti, *Dependence of the Inner DM Profile on the Halo Mass*, Mon. Not. Roy. Astron. Soc. **344** (2003) 1237, [astro-ph/0212146]. 61

[132] J. J. Binney and N. W. Evans, *Cuspy Dark-Matter Haloes and the Galaxy*, Mon. Not. Roy. Astron. Soc. **327** (2001) L27, [astro-ph/0108505]. 62

[133] W. J. G. de Blok, S. S. McGaugh, A. Bosma, and V. C. Rubin, *Mass Density Profiles of LSB Galaxies*, Astrophys. J. **552** (2001) L23–126, [astro-ph/0103102]. 62

[134] R. Kuzio de Naray, S. S. McGaugh, W. J. G. de Blok, and A. Bosma, *High Resolution Optical Velocity Fields of 11 Low Surface Brightness Galaxies*, Astrophys. J. Suppl. **165** (2006) 461–479, [astro-ph/0604576]. 62

[135] J. F. Navarro, C. S. Frenk, and S. D. M. White, *A Universal Density Profile from Hierarchical Clustering*, Astrophys. J. **490** (1997) 493–508, [astro-ph/9611107]. 62

[136] **Particle Data Group** Collaboration, C. Amsler *et. al.*, *Review of particle physics*, Phys. Lett. **B667** (2008) 1. 63, 67

[137] E. I. Gates, G. Gyuk, and M. S. Turner, *The Local halo density*, Astrophys. J. **449** (1995) L123–L126, [astro-ph/9505039]. 63, 67

[138] F. James and M. Roos, *Minuit: A System for Function Minimization and Analysis of the Parameter Errors and Correlations*, Comput. Phys. Commun. **10** (1975) 343–367. 64

[139] F.-S. Ling, *Is the Dark Disc contribution to Dark Matter Signals important ?*, 0911.2321. 67, 71

[140] C. W. Purcell, J. S. Bullock, and M. Kaplinghat, *The Dark Disk of the Milky Way*, Astrophys. J. **703** (2009) 2275–2284, [0906.5348]. 69

[141] M. Bellazzini *et. al.*, *Detection of the Canis Major galaxy at $(l:b) = (244°\text{-}8°)$ and in the background of Galactic open clusters*, Mon. Not. Roy. Astron. Soc. **354** (2004) 1263–1278, [astro-ph/0311119]. 69

[142] M. Bellazzini *et. al.*, *The core of the Canis Major galaxy as traced by Red Clump stars*, Mon. Not. Roy. Astron. Soc. **366** (2006) 865–883, [astro-ph/0504494]. 69

[143] N. F. Martin *et. al.*, *A radial velocity survey of low Galactic latitude structures: I. Kinematics of the Canis Major dwarf galaxy*, Mon. Not. Roy. Astron. Soc. **362** (2005) 906–914, [astro-ph/0503705]. 69

[144] **SDSS** Collaboration, J. Penarrubia *et. al.*, *A comprehensive model for the Monoceros tidal stream*, Astrophys. J. **626** (2005) 128–144, [astro-ph/0410448]. 74, 111

[145] E. B. Hughes *et. al.*, *CHARACTERISTICS OF THE TELESCOPE FOR HIGH-ENERGY GAMMA-RAY ASTRONOMY SELECTED FOR DEFINITION STUDIES ON THE GAMMA-RAY OBSERVATORY. (TALK)*, . IEEE Trans.Nucl.Sci.27 (1980) 364-369.(See Conference Index). 75

[146] G. e. a. Kanbach, *The project egret (energetic gamma-ray experiment telescope) on nasa's gamma-ray observatory (gro)*, Space Sci. Rev. **49** (1988) 69–84. 75

[147] D. J. e. a. Thomson, *Calibration of the energetic gamma-ray experiment telescope (egret) for the compton gamma-ray observatory*, Astrophys. J. **86** (1993) 629–656. 75

[148] EGRET public data archive
ftp://cossc.gsfc.nasa.gov/compton/data/egret/
Date: January 18th, 2010. 76

[149] CGRO Science Support Center webpage
http://heasarc.gsfc.nasa.gov/docs/cgro/egret/
Date: January 18th, 2010. 77

[150] V. Dokuchaev. private communication. 80

[151] I. Gebauer, *An Anisotropic Model for Galactic Cosmic Ray Transport and its Implication for Indirect Dark Matter Searches*. PhD thesis, Karlsruher Institut of Technology, 2010. 99, 104, 112

[152] **GLAST** Collaboration, A. Moiseev, J. Norris, J. Ormes, S. Ritz, and D. Thompson, *Anticoincidence detector for GLAST*, Bull. Am. Astron. Soc. **31** (1999) 721, [astro-ph/9912138]. 101

[153] A. A. Moiseev et. al., *The anti-coincidence detector for the GLAST large area telescope*, Astropart. Phys. **27** (2007) 339–358, [astro-ph/0702581]. 101

[154] Fermi Science Support Center webpage
http://fermi.gsfc.nasa.gov/ssc/data/access/
Date: January 18th, 2010. 102

[155] T. A. Porter and f. t. F. L. Collaboration, *Fermi LAT Measurements of the Diffuse Gamma-Ray Emission at Intermediate Galactic Latitudes*, 0907.0294. 102

[156] Diffuse and Molecular Clouds Working Group Fermi/LAT
http://fermi.gsfc.nasa.gov/ssc/data/access/lat/BackgroundModels.html
Date: January 18th, 2010. 103, 104

[157] P. M. W. Kalberla et. al., *The Leiden/Argentine/Bonn (LAB) Survey of Galactic HI: Final data release of the combined LDS and IAR surveys with improved stray-radiation corrections*, astro-ph/0504140. 104

[158] T. M. Dame, D. Hartmann, and P. Thaddeus, *The Milky Way in Molecular Clouds: A New Complete CO Survey*, Astrophys. J. **547** (2001) 792–813, [astro-ph/0009217]. 104

[159] W. B. Atwood. private communication. 106

[160] B. Alpat, *Alpha Magnetic Spectrometer (AMS02) experiment on the International Space Station (ISS)*, astro-ph/0308487. 109

[161] M. Herold, *Untersuchung der Neutralino-Annihilationen im Rahmen des AMS-02 Experiments*, Master's thesis, Insitut für Experimentelle Kernphysik, Universität Karlsruhe, 2004. in german. 109

[162] M. Honma and Y. Sofue, *Rotation Curve of the Galaxy*, Publ. of the Astronomical Society of Japan **49** (1997) 453–460. 119, 120

Acknowledgements

I would like to thank everyone I was working with and who supported me during my research time in Karlsruhe.

Special thanks go to my supervisor Prof. Dr. Wim de Boer for the possibility to work on this thesis and for his advice and interest.

I want to thank Prof. Dr. Guido Drexlin for acting as second referee of this thesis. I also thank Prof. Dr. Thomas Müller for his work as head of the insitute.

Furthermore, I would like to thank:

Dr. Christian Sander for his help and guidance in the early phase of my work. His knowledge and understanding of cosmic connections was essential to form the basis of my work.

The members of the administration team who keep our computers up to date and provide as much computing power and storage as possible.

Dr. Iris Gebauer, Dr. Martin Heck, Dipl. Phys. Jeaninne Deger-Glaeser and Dipl. Phys. Florian Enders for creating a pleasant atmosphere in our office.

Dipl. Phys. Jasmin Gruschke for her unresting dedication to correct this work. Also many thanks to Dr. Christian Sander, Dr. Iris Gebauer, Dr. Valery Zhukov, Dr. Christopher Jung, Simon Kunz, Conny Beskidt and Anna Owen for proofreading this thesis.

Dipl. Phys. Eva Ziebarth for listening to the rehearsals of my examination talk so many times.

Anna Owen for her patience and her appreciation. I thank her for accepting all the lacks in my free time, especially in the late phase of my work. I also thank my parents for supporting me all the time. Without their backing this thesis would not have been possible.

Die VDM Verlagsservicegesellschaft sucht für wissenschaftliche Verlage abgeschlossene und herausragende

Dissertationen, Habilitationen, Diplomarbeiten, Master Theses, Magisterarbeiten usw.

für die kostenlose Publikation als Fachbuch.

Sie verfügen über eine Arbeit, die hohen inhaltlichen und formalen Ansprüchen genügt, und haben Interesse an einer honorarvergüteten Publikation?

Dann senden Sie bitte erste Informationen über sich und Ihre Arbeit per Email an *info@vdm-vsg.de*.

Sie erhalten kurzfristig unser Feedback!

VDM Verlagsservicegesellschaft mbH
Dudweiler Landstr. 99   Telefon +49 681 3720 174
D - 66123 Saarbrücken   Fax   +49 681 3720 1749
www.vdm-vsg.de

Die VDM Verlagsservicegesellschaft mbH vertritt

Printed by Books on Demand GmbH, Norderstedt / Germany